Hard Math

for

Elementary School:

Answer Key for Workbook

Glenn Ellisc

D1247318

ISBN-10:   1-4848-5142-0
ISBN-13:   978-1-4848-5142-5

Introduction

Most books in the *Hard Math* series have long introductions that tell you something about the book, how to use it, the philosophy behind it, etc. This one doesn't. It's just the answer key for *Hard Math for Elementary School: Workbook*. So basically it just does what you would expect: it gives the answers to the problems in the *Workbook*.

I reproduced the worksheets because I think it makes checking one's answers easier. At the bottom of many worksheets I added brief explanations of how to do some of the harder problems.

I hope that I got all of the answers right. But it's hard to avoid making any mistakes when working on a project this big, so please accept my apologies if you find things that are wrong. I'll try to fix any mistakes in some future edition. And until I do, you could think of the mistakes as a feature: they help show that the problems really are hard and students should feel proud of how many they are able to get.

1.1 Basic Addition with Carrying

1. Word Problem: If Ron has 2127 pieces of candy and Harry has 5374 pieces of candy, how much candy do they have in total?

   > 7,501 pieces

2. Solve the following problems.

```
  11
 128        34          64         894
+ 284      + 28        + 48       + 689
-----      ----        ----       -----
 412        62         112        1583


 102       1111       20102        357
+ 9898     + 7898     + 3256      + 79898
------     ------     -------     -------
10000       9009      23358       80255
```

3. Set up the following problems with one number above the other and then use addition with carrying to find the answers.

   194+329 = 523   46+98 = 144   60+79 = 139

   ```
     11
    194
   +329
   ----
    523
   ```

   987+568 = 1555   695+1234 = 1929 1111+2568 = 3679

4. Challenge: Emma added a 3 digit number ABC to 782 using addition with carrying. She carried ones in the places I've shown. What is the largest possible answer that she could have gotten? (Assume that A, B, and C are all different digits.)

   ```
    1  1
     782              1690
   + ABC
   ```

   Notes on 4: To make the sum as large as possible you want A as large as possible. If A=9 is possible that will be the solution, so try A=9. In the tens column 8 + B + 1 does not result in anything being carried to the hundreds column, so we must have B=0. In the ones column 2 + C results in a 1 being carried to the tens column, so we must have C=8 or 9. We used 9 already so use C=8. 782 + 908 = 1690 .

HMES

1.2 Working with Bigger Numbers

1. Word Problem: Ingrid made 5768425392 Zimbawean dollars from selling squid and 21736808 Zimbabwean dollars from selling shark. How much money did she make in all?

    5,790,162,200 Zimbabwean dollars

2. Insert commas into the following numbers.

    4, 7 8 9, 5 6 2   8, 9 5 6, 7 7 6, 2 1 7   7, 5 2 1, 4 8 9, 6 3 2

3. Write the following numbers out in words.

    4789562   Four million, seven hundred eighty nine thousand, five hundred sixty two

    7521489632   Seven billion, five hundred twenty one million, four hundred eighty nine thousand, six hundred thirty two

4. Arrange the following addition problems so that one number is above the other and solve.

    4789562 + 5562842 = 10,352,404   1243217 + 875624 = 2,118,841

    ```
      1 111  11
      4,789,562
    + 5,562,842
     10,352,404
    ```

5. The Earth's orbit around the Sun is not a perfect circle. When the Earth is closest to the Sun in early January it is about 91,402,640 miles from the Sun. When the Earth is farthest from the Sun in early July it is 3,106,820 miles farther from the Sun. How far from the Sun is the Earth at its farthest point?

    94,509,460 miles

6. Challenge: What is two hundred fifty six trillion, five hundred twenty nine million, six hundred twenty nine thousand, five hundred three plus twenty nine quintillion, one hundred eighty six quadrillion, two hundred ninety nine trillion, four hundred seven million, sixty seven thousand, one hundred ninety nine?

    29,186,555,000,936,696,702

    Notes on 6: Just line up the numbers carefully and do addition with carrying. Be careful to put zeros where needed. The names for each three digit block after the millions (in US English not in the UK) are billion, trillion, quadrillion, and quintillion.

    ```
        256,000,529,629,503
    + 29,186,299,000,407,067,199
    ```

1.3 Working with Decimals

1. Word Problem: Artemis bought $1043.78 of cameras to mount around his house and $228.33 of motion detectors. How much did he spend in total on cameras and motion detectors?

$1272.11

2. Insert commas into the following numbers.

   1 3,8 2 8.2 3 2   1, 2 9 5, 8 6 2.4 8 6 9   7 8 9, 6 5 8.4 7 8 5

3. Write the following numbers out in words.

   13828.232   thirteen thousand, eight hundred twenty eight and two hundred thirty two thousandths

   752.15   seven hundred fifty two and fifteen hundredths

   5000964.263 five million, nine hundred sixty four and two hundred sixty three thousandths

4. Arrange the following addition problems so that one number is above the other and use addition with carrying to solve.

   $12.48 + 78.15 = 90.63$ $124.6 + 35.35 = 159.95$

   $337.78 + 23.441 = 361.221$ $87654321.01234 + 123.45 =$
   $87,654,444.46234$

5. Challenge: At Ralphs's Sandwich Shop a turkey sandwich costs $5.56, a roast beef sandwich costs $7.72, a chicken salad sandwich costs $6.21, a prosciutto and mozzarella sandwich costs $9.34, and a peanut butter and jelly sandwich costs $3.75 Megan's mother buys three sandwiches that together cost exactly $21. Which sandwiches did she buy?

   Two roast beef sandwiches and one turkey sandwich

Notes on 5: One approach is to focus on trying to get the sum to be an even number of dollars. The solution must involve at least two different sandwiches, so you can organize your work around a list of the ten possible pairs written in some logical order, e.g turkey+rb, turkey+cs, turkey+pm, etc. Then, for each pair add the cents and see if there is a third sandwich that makes a whole number of dollars. The first pair in this order has turkey+rb = 56c + 72c = 28c, so a sandwich with a price ending in 72 will give a whole number of dollars. This suggests the two roast-beef, one turkey combination, which we find does work by adding 5.56 + 7.72 + 7.72.

HMES   © 2013 Glenn Ellison

1.4 Adding Lists of Numbers with a Pattern

1. Word Problem: Every year, for her birthday, Abigail gets five times her age in dollars. So, for example, when she turned 3, she got \$15, and when she turned 20, she got \$100. If Abigail is 20, how much money has she gotten so far?

$$\boxed{\$ 1,050}$$

2. Add the following.

10	22	798
12	33	8079
14	44	9807
+16	+55	+7980
52	154	26664

21		13254
32	100	32541
43	101	25413
54	110	54132
+65	+111	+41325
215	422	166665

3. Find the sum of the positive even integers from 30 to 58, i.e. find 30 + 32 + 34 + 36 + ... + 58.

$$\boxed{660}$$

4. Find the sum of all positive six-digit whole numbers that have one digit equal to 4 and the rest of their digits equal to 1. For example, two such numbers are 111,114 and 141,111.

$$\boxed{999,999}$$

5. Challenge: List all three digit numbers that have only 1's and 2's as digits. What is the sum of all of these numbers?

$$\boxed{111,\ 112,\ 121,\ 122,\ 211,\ 212,\ 221,\ 222}$$

$$\boxed{\text{The sum is } 1,332.}$$

Notes on 5: It is always a good idea to put your lists in a clear, logical order. Here, I organized the numbers from smallest to largest. To add the numbers you can write them as an addition with carrying problem. Each column has 4 1's and 4 2's for a sum of 12. The ones that are carried make the answer $\boxed{1332}$.

Name _____ **Answer Key** _____

2.1 Our Base Ten Number System

1. Word Problem: Chiara baked 239 cookies. She wants to put them in as few boxes as possible, but must fill each box that she uses completely. If a large box holds 100 cookies, a medium box holds 10 cookies, and a small box holds 1 cookie, how many boxes of each kind should Chiara use?

    Two large boxes, three medium boxes, and nine small boxes

2. Write out what the following numbers mean.

    10   *1 ten, 0 ones*

    43   *4 tens and 3 ones*

    4,209   *4 thousands, two hundreds, and 9 ones*

3. Draw boxes illustrating what the following numbers mean.

    54 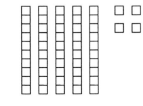   27

    100 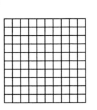   129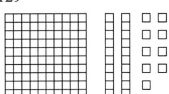

4. In the numbers below circle the digit that is in the hundreds place and underline the digit that is in the ones place.

    (1)2**8**   (3)2**7**   24(3)8**7**   123(4)5**6**   2,000(0)0**1**

5. Challenge: Suppose that you wrote a giant number by writing all the numbers from 1 to 100 in order without any spaces between them. The number would look like 1234567891011…99100. What digit would be in the trillions place?

    9

    Notes on 5: Forget about the start of the number and just work back from the end. One trillion is one followed by 12 zeros, so the trillions place is the 13th digit from the end. The digits from right to left are 0, 0, 1, 9, 9, 8, 9, 7, 9, 6, 9, 5, 9. So the answer is 9. An even quicker way is to notice that the 13th digit from the end will be the 10's digit of some number in the 90's.

2.2 The Base Eight Number System

1. Word Problem: Tian bakes some cookies for the bake sale. She fills 3 medium boxes, each of which holds 8 cookies, and 7 small boxes, each of which holds 1 cookie. How would you write the number of cookies that she baked in base 8?

    $$37_{(8)}$$

2. Write out what the following numbers mean. (For example, $34_{(8)}$ means 3 eights and 4 ones.)

    $43_{(8)}$ 4 eights and 3 ones

    $13_{(8)}$ 1 eight and 3 ones

    $34_{(8)}$ 3 eights and 4 ones

    $77_{(8)}$ 7 eights and 7 ones

3. Draw boxes illustrating what the following numbers mean.

    $57_{(8)}$  $22_{(8)}$

    $45_{(8)}$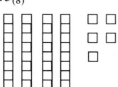

4. Challenge: Bart Simpson is counting his $54_{(8)}$ pranks on his fingers. He puts up a finger each time he counts a prank. And whenever all eight of his fingers are up, he puts them all back down. How many times does he put down all his fingers? How many fingers does he have up at the end?

    He puts them all down 5 times and has 4 up at the end.

    Notes on 4: When counting on your fingers the number of times all fingers go down is just the number of tens. For the Simpsons it is the number of eights. And the number of fingers left up at the end is the number of ones.

2.3 Counting in Base 8

1. Word Problem: If Maggie was $34_{(8)}$ days old yesterday, how many days old is she today? Give your answer in base 8.

$$35_{(8)}$$

2. Fill in the missing numbers. (Remember that these numbers are in base 8, not base 10.)

$0_{(8)}, 1_{(8)}, 2_{(8)}, \underline{3}_{(8)}, 4_{(8)}, \underline{5}_{(8)}, 6_{(8)}, 7_{(8)}, \underline{10}_{(8)}, 11_{(8)}, \underline{12}_{(8)}, 13_{(8)}, 14_{(8)},$

$\underline{15}_{(8)}, 16_{(8)}, \underline{17}_{(8)}, 20_{(8)}, 21_{(8)}, 22_{(8)}, \underline{23}_{(8)}, 24_{(8)}, \underline{25}_{(8)}, 26_{(8)},$

$\underline{27}_{(8)}, 30_{(8)}, 31_{(8)}, 32_{(8)}, \underline{33}_{(8)}, 34_{(8)}, \underline{35}_{(8)}, 36_{(8)}, \underline{37}_{(8)}, \underline{40}_{(8)}$

$\underline{312}_{(8)}, \underline{313}_{(8)}, \underline{314}_{(8)}, \underline{315}_{(8)}, \underline{316}_{(8)}, \underline{317}_{(8)},$

$\underline{320}_{(8)}, \underline{321}_{(8)}, \underline{322}_{(8)}, \underline{323}_{(8)}, 324_{(8)}$

$\underline{6676}_{(8)}, \underline{6677}_{(8)}, 6700_{(8)}, \underline{6701}_{(8)}, \underline{6702}_{(8)}, 6703_{(8)}, 6704_{(8)}, 6705_{(8)},$

$\underline{6706}_{(8)}, \underline{6707}_{(8)}, \underline{6710}_{(8)}, 6711_{(8)}, \underline{6712}_{(8)}, \underline{6713}_{(8)}$

3. What number is three more than $55_{(8)}$?

$$60_{(8)}$$

4. What number is half way in between $57_{(8)}$ and $65_{(8)}$?

$$62_{(8)}$$

5. Challenge: What base 8 number comes seven numbers after the number that is one less than the number that is three more than $57_{(8)}$?

$$70_{(8)}$$

Notes on 5: One way to do this is to combine "seven after" "one less than" "three more" to 7-1+3 = 9 more. Another way is just to work backwards keeping everything in base 8. Three more than $57_{(8)}$ is $62_{(8)}$. One less than this is $61_{(8)}$. And seven more than this is $70_{(8)}$.

HMES   © 2013 Glenn Ellison

2.4 Bigger Numbers in Base 8: The Sixty-Fours Place

1. Word Problem: A chessboard is an 8 by 8 grid with alternating white and black squares. Lucia puts two pennies on each square. How many pennies are on the board in total?

   128

2. Write the meanings of the following base 8 numbers.

   $100_{(8)}$ *1 sixty four, 0 eights, 0 ones*

   $107_{(8)}$ *1 sixty four, 0 eights, 7 ones*

   $262_{(8)}$ *2 sixty fours, 6 eights, 2 ones*

   $1,010_{(8)}$ *1 five hundred twelve, 0 sixty fours, 1 eight, 0 ones*

3. In the numbers below circle the digit in the 64's place and underline the digit in the one's place. Put an arrow pointing toward any digit in the 512's place.

   $40\underline{0}_{(8)}$ $(1)0\underline{7}_{(8)}$ $(2)62_{(8)}$ $3,(1)0\underline{7}_{(8)}$ $21(2)6\underline{2}_{(8)}$

4. Put the following base 8 numbers in order from smallest to biggest: $430_{(8)}$, $257_{(8)}$, $103_{(8)}$, and $1,104_{(8)}$?

   $103_{(8)}, 257_{(8)}, 430_{(8)}, 1,104_{(8)}$

5. Sonia has 3 boyfriends. On Valentine's Day each of them gives her a box of 64 chocolates. How would you write the number of chocolates that she gets in base 8?

   $300_{(8)}$

6. Challenge: In the Simpsons' elementary school (which uses base 8) the teacher asks the students to write down the smallest whole number that uses every digit exactly once. What is the correct answer?

   $10,234,567_{(8)}$

   Notes on 6: In base 8 there are eight digits: 0, 1, 2, 3, 4, 5, 6, and 7. To make the smallest number possible, put the smaller digits first. But you need to start with a 1 instead of a 0 because a number can't start with a zero.

2.5 Addition in Base 8

1. Word Problem: Lisa Simpson buys $34_{(8)}$ Olympic pins, steals $12_{(8)}$, trades for $20_{(8)}$, and is given 1. How many Olympic pins does she get in total? Give your answer in base 8.

$$\boxed{67_{(8)}}$$

2. Add the following base 8 numbers.

$2_{(8)}$	$12_{(8)}$	$15_{(8)}$	$23_{(8)}$
$+3_{(8)}$	$+25_{(8)}$	$+31_{(8)}$	$+43_{(8)}$
$5_{(8)}$	$37_{(8)}$	$46_{(8)}$	$66_{(8)}$

$23_{(8)}$	$10_{(8)}$	$4_{(8)}$	$62_{(8)}$
$+52_{(8)}$	$+57_{(8)}$	$+63_{(8)}$	$+102_{(8)}$
$75_{(8)}$	$67_{(8)}$	$67_{(8)}$	$164_{(8)}$

3. Rewrite the following addition problems so that one number is on top of the other and add.

$2_{(8)}+5_{(8)}=\boxed{7_{(8)}}$ $4_{(8)}+21_{(8)}=\boxed{25_{(8)}}$ $15_{(8)}+22_{(8)}=\boxed{37_{(8)}}$ $20_{(8)}+27_{(8)}=\boxed{47_{(8)}}$

$$\begin{array}{r} 2 \\ +5 \\ \hline 7 \end{array}$$

$60_{(8)}+115_{(8)}=\boxed{175_{(8)}}$ $63_{(8)}+204_{(8)}=\boxed{267_{(8)}}$ $70_{(8)}+201_{(8)}=\boxed{271_{(8)}}$ $73_{(8)}+300_{(8)}=\boxed{373_{(8)}}$

4. Making Connections: What is $123_{(8)} + 231_{(8)} + 312_{(8)}$? Give your answer in base 8.

$$\boxed{666_{(8)}}$$

5. Challenge: Suppose that people in the Simpsons' world watched a TV show in which people had only 6 fingers. Suppose that their version of *Hard Math* talked about this show and taught them about base 6 math. What would they answer if they were asked to find $123_{(6)} + 231_{(6)} + 312_{(6)}$?

$$\boxed{1110_{(6)}}$$

Notes on 4: The connection is to section 1.4. Using addition with carrying all three columns add to 6. On 5: The three columns again add to 6. But in base 6 this is a zero with a one carried to the next place value.

2.6 Addition with Carrying in Base 8

1. Word Problem: Comic Book Guy has $15_{(8)}$ Catwoman comics, $23_{(8)}$ Spiderman comics, and $72_{(8)}$ Batman comic books in his private collection. How many comics does he have in his private collection in all?

   $\boxed{132_{(8)}}$

2. Add the following numbers. Give your answers in base 8.

   $2_{(8)}$ $15_{(8)}$ $15_{(8)}$ $27_{(8)}$ $36_{(8)}$
   $+7_{(8)}$ $+25_{(8)}$ $+33_{(8)}$ $+44_{(8)}$ $+25_{(8)}$
   $11_{(8)}$ $42_{(8)}$ $50_{(8)}$ $73_{(8)}$ $63_{(8)}$

   $41_{(8)}$ $52_{(8)}$ $57_{(8)}$ $62_{(8)}$ $164_{(8)}$
   $+50_{(8)}$ $+53_{(8)}$ $+63_{(8)}$ $+66_{(8)}$ $+\ 72_{(8)}$
   $111_{(8)}$ $125_{(8)}$ $142_{(8)}$ $150_{(8)}$ $256_{(8)}$

3. Rewrite the following addition problems so that one number is on top of the other and add. Give your answers in base 8.

   $2_{(8)}+11_{(8)}=$ $4_{(8)}+24_{(8)}=$ $15_{(8)}+27_{(8)}=$ $17_{(8)}+27_{(8)}=$
   $\boxed{13_{(8)}}$ $\boxed{30_{(8)}}$ $\boxed{44_{(8)}}$ $\boxed{46_{(8)}}$

   $60_{(8)}+165_{(8)}=$ $63_{(8)}+207_{(8)}=$ $72_{(8)}+777_{(8)}=$ $13_{(8)}+777_{(8)}=$
   $\boxed{245_{(8)}}$ $\boxed{272_{(8)}}$ $\boxed{1,071_{(8)}}$ $\boxed{1,012_{(8)}}$

4. Making Connections: What is $31406524_{(8)} + 203456314_{(8)}$? Give your answer as a base 8 number

   $\boxed{235{,}065{,}040_{(8)}}$

5. Challenge: What is the sum of all 3 digit base 8 numbers with each digit being either 1 or 2? Give your answer in base 8.

   $\boxed{1554_{(8)}}$

Notes on 4: The connection is to section 1.2. When adding big numbers it's useful to put in the commas to help line things up. On 5: One way is to write down all 8 numbers in a logical order – 111, 112, 121, 122, 211, 212, 221, 222 – and then use addition with carrying. But one doesn't really need to write out the list. If you know that there are eight such numbers and four will have 1's and four will have 2's in each place, then it follows that each place value in the addition problem will sum to $12 = 14_{(8)}$. Carrying the 1's gives $1554_{(8)}$ as the answer.

2.7 We Need a Translator: Converting Base 8 Numbers to Base 10

1. Word Problem: Bart and Lisa have found a machine that goes from the Simpsons' world to the real world. Professor Frink, who is $42_{(8)}$ years old, is the first to go through. However, when he gets to the other side, he finds that everyone uses base 10. What should he say his age is in base 10?

    $\boxed{34}$

2. Translate the following base 8 numbers to base 10.

    $53_{(8)}$

    $5 \times 8 + 3 = 43$

    $14_{(8)}$

    $1 \times 8 + 4 = \boxed{12}$

    $64_{(8)}$

    $6 \times 8 + 4 = \boxed{52}$

    $102_{(8)}$

    $1 \times 64 + 2 = \boxed{66}$

    $162_{(8)}$

    $(1 \times 64) + (6 \times 8)$
    $+ 3 = \boxed{115}$

    $1145_{(8)}$

    $(1 \times 512) + (1 \times 64)$
    $+ (4 \times 8) + 5 = \boxed{613}$

3. Suppose that *Hard Math* is offered for sale in both base 8 and base 10 pricing. You have a choice of paying either $6000_{(8)}$ or $2995_{(10)}$ cents. Which is a better deal?

    $\boxed{2995_{(10)} \text{ is a better deal.}}$

4. Making Connections: What is $10_{(8)} + 20_{(8)} + 30_{(8)} + 40_{(8)} + 50_{(8)}$? Write your answer as a base 10 number.

    $\boxed{120}$

5. Challenge: The numbers 1 through 7 all have the property that the sum of their digits is the same regardless of whether you write them in base 8 or base 10. What is the next smallest number with this property?

    $\boxed{70_{(10)} = 106_{(8)}}$

Notes on 3: The place value after 64 in base 8 is the 512's place. So $6000_{(8)} = 6 \times 512 = 3072$. On 4: The connection is to section 1.4. $1+2+3+4+5=15$, so we can think of the sum as 15 eights and 0 ones. $15 \times 8 = 120$. On 5: This is hard. The best way to solve it in elementary school is probably just to be organized in trying numbers counting up from 8. 8 and 9 clearly don't work because the base 10 sum is 7 larger. Numbers from 10 to 15 don't work because the sum of the digits is 2 smaller in base 10. Numbers from 16-19 don't work because the base 10 sum is 5 bigger. Then the base 10 sum is 4 smaller for 20-23 and so on. (Note that the difference in the two sums only changes when you cross a multiple of 8 or 10 so you only need to check 16, 20, 24, 30, 32, 40, 48, 50, 56, 60, 64, …) Each block similarly doesn't work until you come to 70 which does work (as does 71).

2.8 What Would the Simpsons Call Numbers?

1. Suppose that the English names for the base 10 numbers 10, 20, 30, …, 90 all followed the most common pattern. Which numbers would we say differently and how would we say them?

   10 – "onety"   20 – "twoty"   30 – "threety"   50 – "fivety"

2. In Chinese the names for numbers are more logical. After "ten" they just call the next numbers "ten one", "ten two", …, "ten nine". And then instead of making up words like "twenty" they say "two ten" for "twenty", "two ten one" for twenty one, and so on. Of course, they don't exactly say this. They say the Chinese words for each digit: 1="yi", 2="er", 3="san", 4="si", 5="wu", 6="liu", 7="qi", 8="ba", 9 = "jiu", 10="shi". How would you say the following numbers in Chinese?

   13 – shi san   32 – san shi er

   27 – er shi qi   49 – si shi jiu

   52 – wu shi er   88 – ba shi ba

3. Suppose that you were making up names for base 8 numbers what numbers would you make up for the numbers below?

   $20_{(8)}$ – $30_{(8)}$ – $40_{(8)}$ –
   $50_{(8)}$ – $60_{(8)}$ – $70_{(8)}$ –
   $100_{(8)}$ –

   Answers can vary

   What names would you then give to

   $57_{(8)}$ –

   $237_{(8)}$ –

   Answers can vary, but should be logically related to the first answers.

4. When we write big base 10 numbers in English we put commas and give new names every three digits: thousand, million, billion, trillion, … But it didn't have to be this way. In Chinese, ten thousands are more important: there's a special word for "ten thousand" and you say things kind of like "three thousand fourteen ten thousands" for 30140000. If you were making up a way to write and say bigger numbers in base 8, where would you put the commas and what names would you give to:

   $10_{(8)}$ – $100_{(8)}$ – $1000_{(8)}$ –

   $10000_{(8)}$ – $1000000_{(8)}$ – $1000000000_{(8)}$ –

   Notes on 4: Answers can vary, but the structure should line up with where the commas are placed. For example, if commas are placed every 4 digits then $1,0000_{(8)}$ and $1,0000,0000_{(8)}$ would each get a new one-word name, $1000000_{(8)}$ would have the comma placed as $100,0000_{(8)}$, and it would be called the name for $100_{(8)}$ followed by the name for $1,0000_{(8)}$.

16

Hard Math **Worksheets**

Name _____**Answer Key**_____

3.1 Keep a Running Total and Add the Tens Before the Ones if You Need To

1. Word Problem: River has killed 56 reavers, Jayne has killed 24, Mal has killed 16, Zoe has killed 12, and none of the other crew members have killed any. How many reavers have been killed in all?

    $\boxed{108}$

2. Add the following numbers in your head.

    $25+14+38 = \boxed{77}$ $\qquad\qquad$ $27+50+11 = \boxed{88}$

    $31+42+20 = \boxed{93}$ $\qquad\qquad$ $56+30+23 = \boxed{109}$

    $52+13+69 = \boxed{134}$ $\qquad\qquad$ $28+16+23 = \boxed{67}$

    $79+79+10 = \boxed{168}$ $\qquad\qquad$ $32+15+17+21 = \boxed{85}$

    $21+54+62+78 = \boxed{215}$ $\qquad\qquad$ $64+35+12+81 = \boxed{192}$

    $95+15+42+36+75 = \boxed{263}$ $\qquad\qquad$ $65+44+32+87+37 = \boxed{265}$

3. Making Connections: Do the following in your head and give your answer in base 8. (It is probably less confusing to add the eights before the ones when there is carrying.)

    $23_{(8)}+41_{(8)}+13_{(8)} = \boxed{77_{(8)}}$ $\qquad\qquad$ $35_{(8)}+33_{(8)}+23_{(8)} = \boxed{113_{(8)}}$

4. Challenge: Working in your head find the sum of all two digit positive whole numbers that start with a 2.

    $\boxed{245}$

Notes on 3: The connection is to section 2.6. I find it easier to say the numbers as digits, e.g. two-three, six-three, six-four, seven-four, seven-seven. On 4: The 2-digit numbers that start with a 2 are 20, 21, 22, 23, 24, 25, 26, 27, 28, 29. One way to add them is to first add the 20's to get 200 then count up 201, 203, 206, …, 245. Or you can just add the whole numbers one at a time: 20, 41, 63, 86, (106-110), 135, (155-161), 188, (208-216), (236-245).

HMES © 2013 Glenn Ellison

3.2 Look for an Ordering that Makes the Addition Problem Easier

1. Word Problem: On Monday, Daenarys eats 86 figs. On Tuesday, she eats 78 figs. On Wednesday, she only eats 14 figs. How many figs does she eat in total on those three days? $\boxed{178}$

2. Rewrite the following addition problems in an order that makes them easier to solve and then solve.

    $78+56+22 = (78+22)+56 = 156$

    $64+75+25 = 64 + (75 + 25) = 164$

    $11+95+89 = (11 + 89) + 95 = 195$

    $60+27+39 = (60 + 39) + 27 = 126$

    $56+999+1 = 56 + (999 + 1) = 1056$

3. Solve the following problems.

    $64+75+13+12 = 164$

    $12+16+88 = \boxed{116}$

    $95+90+5+7 = \boxed{197}$

    $57+11+2+89+42 = \boxed{201}$

4. Making Connections: Base 8 addition problems are easier if you can find pairs like $50_{(8)}$ and $30_{(8)}$ or $62_{(8)}$ and $18_{(8)}$ that add up to $100_{(8)}$. Use this strategy to add the numbers below in your head. Give your answer in base 8.

    $23_{(8)} + 42_{(8)} + 36_{(8)} = \boxed{123_{(8)}}$

5. Challenge: Working in your head find $82.65 + 32.77 + 18.23 + 26.22$.

    $\boxed{159.87}$

    Notes on 4: Connection to section 2.6. To make it easier note that $42_{(8)} + 36_{(8)} = 100_{(8)}$. On 5: Adding is easier if you do the integer parts before the decimals and reorder each to take advantage of terms that add up to 100, i.e. adding $(82 + 18) + 32 + 26 + (0.77 + 0.23) + 0.65 + 0.22$ by saying 100, 132, 158, 159, 159.65, 159.87.

3.3 Adding Evenly Spaced Numbers

1. Word Problem: Whenever Harvard scores in a football game the male cheerleaders do a number of pushups equal to Harvard's score. For example, if the team has just kicked a field goal to go ahead 17–10 they will do 17 pushups. How many pushups will each male cheerleader have to do in total if the team scores five touchdowns in a game? (Assume they make all extra points and they only do pushups after the extra point so the scores go 7, 14, 21, 28, 35).

   $\boxed{105}$

2. Circle pairs of numbers and draw lines connecting them to show which numbers you'd add to use the working-from-the-outside-in trick in these sums

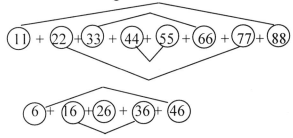

3. Find the sum of each of the sets of numbers below.

   $1 + 3 + 5 + 7 + 9 + 11 = \boxed{36}$

   $5 + 15 + 25 + 35 + 45 = \boxed{125}$

   $11 + 21 + 31 + 41 + 51 + 61 + 71 + 81 + 91 = \boxed{459}$

4. Suppose that you are trying to add $10 + 12 + 14 + \ldots + 40$ by adding from the outside in. What would be the first pair of numbers you'd add? __10 + 40__
   What do they add up to? __50__ What would be the last pair of numbers you'd add? __24 + 26__ How many pairs of numbers are there? __8__ Is there a single number left over in the middle? __No__ What is answer to $10 + 12 + \ldots + 40$?

   $\boxed{400}$

5. Challenge: Find the sum of the first seven odd multiples of 99.

   $\boxed{4851}$

   Notes on 5: You can think of the first seven odd multiples of 99 as being $(100 - 1)$, $(300 - 3)$, $(500 - 5)$, $(700 - 7)$, $(900 - 9)$, $(1100 - 11)$, and $(1300 - 13)$. The first and last terms add up to $1400 - 14$. The 2nd and 6th and 3rd and 5th have the same sum so the sum of all the terms except the middle one is $4200 - 42$. Adding $700 - 7$ gives $4900 - 49 = 4851$.

3.4 Put in Some Extra Additions and Subtractions to Get Round Numbers

1. Word Problem: Mr. Weasley used to have 394 stamps in his collection, but for his birthday Harry gave him 93 stamps and Hermione gave him 112. After his birthday, how many stamps did Mr. Weasley have?

   $$\boxed{599}$$

2. Write out the problem with extra additions and subtractions to get round numbers and solve.

   $199+104+310 = 200 - 1 + 100 + 4 + 300 + 10 = 613$

   $305+498 = 300 + 5 + 500 - 2 = 803$

   $210+399+12 = 200 + 10 + 400 - 1 + 12 = 621$

3. Solve in your head by approximating and then cleaning up.

   $708+298+206 = 1212$ $398+299+409 = \boxed{1106}$

   $307+392+112+235+707 = \boxed{1753}$ $1{,}802+1{,}904+580+404 = \boxed{4690}$

4. Ruitian buys four books. The prices of the four books are $7.95, $12.95, $29.95, and $12.50. How much does he spend in all?

   $$\boxed{\$\ 63.35}$$

5. Making Connections: Rewrite the base 8 sum below as a base 8 problems with extra additions and subtractions. Then find the answer in your head.

   $103_{(8)} + 276_{(8)} + 307_{(8)} = 100_{(8)} + 3_{(8)} + 300_{(8)} - 2_{(8)} + 300_{(8)} + 7_{(8)} = \boxed{710_{(8)}}$

6. Challenge: If you go around the Earth at the equator it is 24,901.55 miles around. The distance from the Earth to the Moon is different every day. Suppose that when you want to go it is 238,900 miles away. How far would you go in total if you drove around the earth at the equator and then drove to the moon and back? (Assume that you had a car that could drive on water and in outer space and do all the additions in your head.)

   $$\boxed{502{,}701.55 \text{ miles}}$$

   Notes on 6: The distance to the moon is $250{,}000 - 11{,}100$ miles. So twice the distance is $500{,}000 - 22{,}200$. And the total is $500{,}000 + (24{,}901.55 - 22{,}200)$.

3.5 Working with Bigger Numbers: Add the Hundreds Then the Rest

1. Word Problem: There are 470 petals on one sunflower, 134 on the next, and 483 on the third. How many petals are there on all three in total?

   $\boxed{1,087}$

2. Solve by adding the hundreds then the rest.

   $768+218+476 = 1462$ $408+541 = \boxed{949}$

   $243+561 = \boxed{804}$ $443+409+527 = \boxed{1,379}$

   $341+714+607+242 = \boxed{1,904}$ $522+230+106+3,141 = \boxed{3,999}$

   $216+1,010+6,789+276+447 = \boxed{8,738}$

3. Kevin's family drove to Maine for their summer vacation. On the way there they broke the driving up over two days driving 284 miles on the first day and 154 on the second. On the way back they drove all 438 miles in one day. How many miles did they drive in all?

   $\boxed{876}$

4. Making Connections: Solve by adding the 64s and then the rest. Give your answer in base 8.

   $112_{(8)} + 231_{(8)} + 1233_{(8)} + 302_{(8)} = \boxed{2100_{(8)}}$

5. Challenge: Solve by adding the hundreds then the rest.

   $10,415 + 2,849 + 52,468 + 31,515 + 4,356 = \boxed{101,603}$

   Notes on 3: It's easier if you realize that you can just double 438 miles. On 4: The connection is to section 2.6. Adding the digits in the 64's place one by one you'd say 1, 3, 15 (pronounced "one five"), 20 ("two-o"). And then going back and adding the one's you'd say 2012, 2043, 2076, 2100. (Again I like to read each number as four digits.) On 5: To do this in my head I'd do it in three passes. First add the thousands, "10, 12, 64, 95, 99," to get 99,000. Then start counting the hundreds, e.g. "4, 12, 16, 21, 24 = 2 thousand 4 hundred so 101,400. Then do the tens and ones "101,415, 101,464, 101,532, 101,547, 101,603".

3.6 A Different Method: Add the Tens and Then the Ones

1. Word Problem: When Julia was at camp, there was an activity where three people had to each introduce themselves to each other. The first person spent 12 minutes talking, the second spent 14 minutes, and Julia spent 53 minutes talking. How much time did they spend introducing themselves in total?

    > *79 minutes or 1 hour 19 minutes*

2. Solve in your head by adding the tens and then the ones.

    78+21+76 = *175*   41+14+83= *138*

    45+50+27= *122*   24+48+22= *94*

    31+14+61+24= *130*   36+49+4+81= *170*

    95+62+47+98+46+21= *369*

3. Making Connections: Add the eights and then the ones in your head to find the answer in base 8:

    $10_{(8)} + 42_{(8)} + 23_{(8)} + 21_{(8)} =$ *116* $_{(8)}$

4. Challenge: The "average" of a set of numbers is what you get if you add all the numbers together and then divide by the number of numbers. For example the average of 40 and 60 is $(40 + 60) \div 2 = 50$ and the average of 20, 52, and 18 is $(20 + 52 + 18) \div 3 = 30$.

    If Mei's scores on the ten tests in a class are 50, 42, 50, 45, 38, 43, 47, 44, 50, and 41, then what is the average of her test scores?

    > *45*

Notes on 3: The connection is to section 2.6. Keeping a running total of the eights sounds like "one, five, seven, one-one." Then continue "one-one-two, one-one-five, one-one-six". On 4: First add the ten numbers. You can add the tens saying "fifty, niney, one-forty, one-eighty, two-ten, …" or as "five, nine, fourteen, eighteen, twenty-one, …, forty-two". (In problems like this sometimes I'll add two at a time, e.g. going straight from 21 to 29 by adding two fours.) From this point count as "422, 427, 435, 438, 445 (or skip straight to 445 when you see the 3 and 7 next to each other), 449, 450. Then divide by ten.

3.7 Reordering in the Second Method: Do the Tens and Ones Separately

1. Word Problem: The Chinese government has decided to fine several people for breaking the law. They fine one farmer 29 yuan, another farmer 76 yuan, a third farmer 31 yuan, and a court official 26 yuan. How much does the government collect from these four people?

   162 yuan

2. Solve in your head by adding the tens and the ones separately. Look to group pairs that sum to 10.

   38+22+74= 134   26+83 = 109

   54+58+22 = 134   93+77+12 = 182

   79+61+46 = 186   13+47+62 = 122

   36+49+64+51= 200   91+17+72+36 = 216

   23+84+69+41+36 = 253   27+14+82+73+28+92 = 316

3. Kate is 47 inches tall. Eva is two inches shorter than Penelope. Penelope is four inches taller than Kate. If the girls go to a circus camp and learn to do a trick where Eva stands on Penelope's head and Kate stands on Eva's head, how far above the ground will the top of Kate's head be?

   147 inches or 12 feet 3 inches

4. Challenge: What is the sum of all positive two-digit whole numbers that do not have a 5 in them? (Hint: Write the numbers in an array and think about pairing digits and adding the tens and then the ones.)

   3920

Notes on 3: Look for things you can figure out instead of trying to find things in order. First, find that Penelope is 51" tall and then that Eva is 49". Add as $47 + (49 + 51)$. On 4: Picture the numbers as an 8×9 matrix as shown below. Add the tens by picking a tens digit in the top half and adding it to the opposite tens digit in the bottom half. 36 pairs each adding to 100 gives 3600. Then, adding the ones by matching a digit in the left half with one on the right half. The 8×8 grid of numbers that don't end in zero gives 32 pairs each adding up to 10. $3600 + 320 = 3920$.

```
10 11 12 13 14   16 17 18 19
20 21 22 23 24   26 27 28 29
 :  :  :  :  :    :  :  :  :
80 81 82 83 84   86 87 88 89
90 91 92 93 94   96 97 98 99
```

4.1 What Makes a Floor Fit Together? Angles

The next two pages of the workbook have outlines of polygons. To do the problems in this chapter you'll want to make lots of polygons just like them out of colored paper. The easiest way is to make several copies of each page on colored paper and then cut out the shapes. Talking someone with access to a copier into copying each onto 6 different colors of construction paper would be ideal. If not, you could cut the shapes out of the workbook, put the cut-out shapes on top of colored paper, trace around them with a pencil, and then cut out the shapes.

1. Word Problem: Kara decided to make a tile floor for her house using triangular tiles. How many tiles will need to meet at the corner that the arrow is pointing to so that there is no empty space between the triangles?

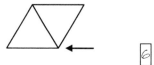

$\boxed{6}$

2. Suppose you start trying to make a tile floor by putting down an octagonal tile and then putting a square tile next to it. Is there a regular polygon that will fit exactly into the space shown?

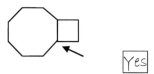

$\boxed{\text{Yes}}$

3. Suppose you try to have three octagons meet at a point. Will they fit together exactly, will there be open space between them, or will they overlap?

$\boxed{\text{They will overlap.}}$

4. Challenge: In the book I showed a failed attempt at tiling a floor with pentagons. Although it is impossible to make a floor with pentagons without leaving spaces, there are ways to fit pentagons together leaving neater, more regular spaces. Find the nicest pentagons-and-spaces tiling pattern you can and draw a picture of it here.

Many answers are possible.

Notes on 2: A regular octagon will fit. In practice it may not fit exactly if the printing of the workbook didn't preserve all angles or if you didn't cut it out correctly. On 4: One way to find nice patterns is to start by putting pentagons along the five sides of one pentagon and then seeing what you can do to make regular spaces. Another approach is to put pentagons so that they border star-shaped or diamond-shaped blank spaces.

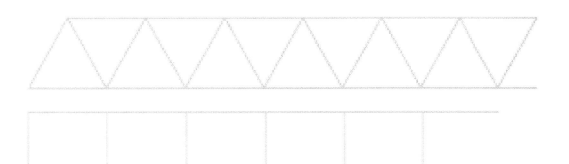

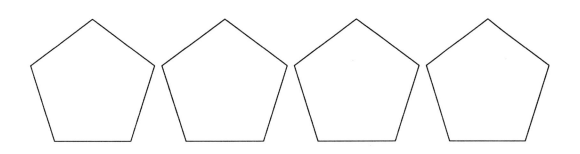

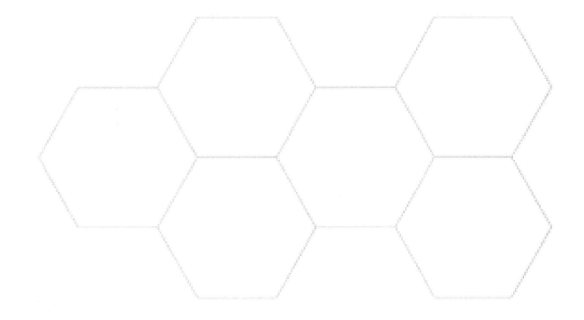

4.2 Measuring Angles

1. Draw lines matching each angle to its measure in degrees.

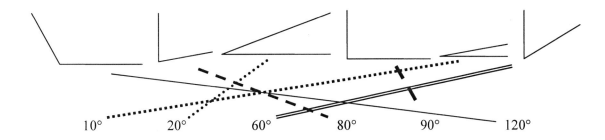

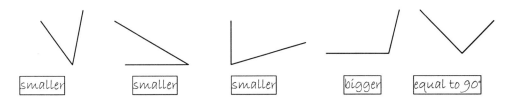

10°   20°   60°   80°   90°   120°

2. For each of the angles below say if it is smaller than 90°, equal to 90°, or larger than 90°.

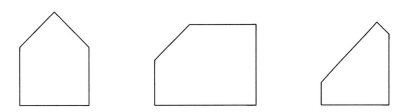

smaller   smaller   smaller   bigger   equal to 90°

3. Draw a four-sided polygon that has right angles at the two bottom corners, an angle less than 90° at the upper left corner, and an angle greater than 90° in the upper right corner.

(Answers can vary)

4. Challenge: Draw a pentagon that has three right angles and the other two angles equal to each other. Then try to draw a second pentagon with this property that is not "similar" to the first. (Two pentagons are similar if one is just a larger version of the other rotated in some way.)

(Answers can vary. Some examples are below.)

29

4.3 Using a Protractor to Measure Angles

1.　Find the measure in degrees of each of the angles below.

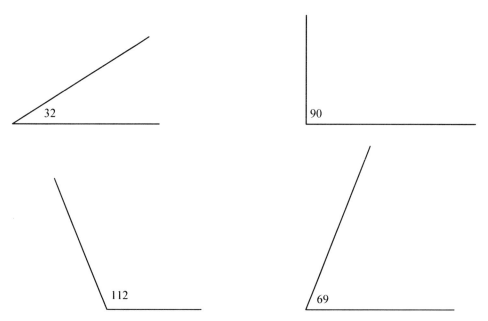

Note: Answers to both questions may vary somewhat both because measurements in the production version may differ slightly and because students might round off answers.

2. Find the measure in degrees of each of the angles in the shapes below. You may need to use a ruler to extend some of the shorter sides to read the measures on your protractor.

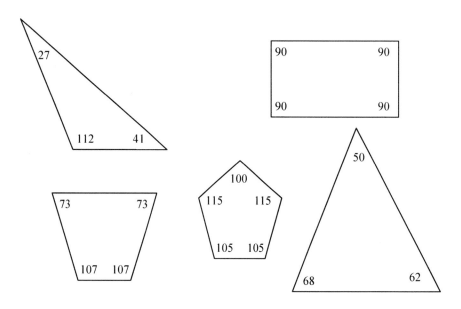

4.4 Regular Polygons

1. Use a protractor to measure one of the angles in the dodecahedron that you cut out of colored paper. (One trick is to trace along the two edges of the angle on scratch paper. You can then use a ruler to make them longer if you need to.) What is the measure in degrees?

   150 degrees

2. You can check whether two of the octagons you cut out are identical by putting one on top of the other and seeing if they line up perfectly. If they do, you can see if they're really regular octagons by rotating the top one, one side at a time, and seeing if it still lines up. Are your two octagons regular octagons?

   Answers can vary depending on how well the printing process preserved the angles and how students cut them out.

3. The book gives a formula for computing the degree measure of each angle in a regular N-gon: it is $180 - (360 \div N)$. Use this formula to compute the degree measure of the angles in a regular N-gon for each value of N listed below.

N	$180 - (360 \div N)$	Drawing of N-gon
3	$180-(360\div3)=180-120 = \underline{60}$	
4	$180-(360\div4)=180-90 = \boxed{90}$	
6	$180-(360\div6)=180-60 = \boxed{120}$	
8	$180-(360\div8)=180-45 = \boxed{135}$	

4. Challenge: One way to draw regular N-gons is to make a perfect circle, draw in N points equally spaced around the circle, and then to use a ruler to connect adjacent points. Use this technique to draw an equilateral triangle, a regular hexagon, and a regular 12-gon.

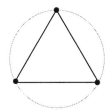  

4.5 What Fits Together? The 360 Rule

1. Using the shapes you cut out, check to see if three triangles and two squares fit together around a point. To check you answer, add up the measures of the five angles to see if they add up to 360°.

> They fit. The angles are 60+60+60+90+90=360.

2. For each of the combinations of shapes listed below add up the measures of the three angles to see if they add up to less than 360, exactly 360, or more than 360. Then use the shapes you cut out to see if the three shapes leave open space, fit exactly, or overlap.

Shapes	Sum of Angles	More, Less, or Equal to 360° ?	Open space, fit, or overlap?
3 squares	90+90+90=270	Less	Open space
3 hexagons	120+120+120=360	Equal	Fit
1 square, 1 hexagon, 1 octagon	90+120+135=345	Less	Open space
1 square, 1 hexagon, 1 dodecagon	90+120+150=360	Equal	Fit
1 hexagon, 2 octagons	120+135+135=390	More	Overlap

3. Challenge: Lillian starts trying to make a tile floor by putting two regular pentagons next two each other. Is there some regular N-gon (not necessarily one of the ones you cut out of paper) that would fit exactly into the open space?

> Yes. A regular decagon (10-gon) fits.

Notes on 3: A regular pentagon has a 108° angle so two pentagons add up to 216°. The open space is 360° – 216° = 144°. The angle in a regular n-gon is 180 – (360)/n. So the angle in a regular 10-gon is 180 – (360/10) = 180 – 36 = 144.

4.6 Using the 360 Rule to Find Regular Tiling Patterns

1. Word Problem: Adolphus claimed that there would be open spaces if you tried to tile a floor using hexagons because the interior angle of a hexagon is 124° and 124° + 124° + 124° = 348° is less than 360°. What is wrong with his argument?

    There are two problems: the interior angle of a hexagon is 120° not 124°; and 124 + 124 + 124 is not 348.

2. Using the colored-paper hexagons that you cut out, see how hexagons can fit together to tile a floor. Draw a picture of a few hexagons to show how you fit them together.

3. What is 90 + 90 + 90 + 90? What tiling pattern is related to this addition problem?

    The sum is 360. The tiling is the standard tiling with square tiles.

4. Color in the triangles in the triangular tiling shown below to make it look like a tiling that uses hexagons. (Hint: You want to color groups of six adjacent triangles using the same color.)

5. Challenge: When Kate went to a palace in Istanbul she saw the tile pattern below on a wall. Can you figure out how to color the triangles in a triangular tiling to make this pattern?

       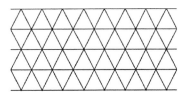

    Notes on 5: The triangular grid needs to be tilted 30 degrees to match the picture. Use a group of 6 triangles to make the hexagons and use 4 triangles to make the triangles.

33

4.7 Semiregular Tilings

1. Find all ways to make 360 as the sum of one 120 and some number of 90's and 60's.
   Organize your work by arranging things in terms of the number of 90's you use: zero, one,
   or two. Not all of them work.

<table>
<tr><td>1 120 & 2 90's</td><td>1 120 & 1 90</td><td>1 120 & 0 90's</td></tr>
<tr><td>120</td><td>120</td><td>120</td></tr>
<tr><td>90</td><td>90</td><td>60</td></tr>
<tr><td>90</td><td></td><td>60</td></tr>
<tr><td>+ 60</td><td>+</td><td>60</td></tr>
<tr><td>360</td><td>Doesn't work</td><td>+ 60</td></tr>
<tr><td></td><td></td><td>360</td></tr>
</table>

2. One addition problem that adds up to 360 is 120 + 120 + 60 + 60 = 360. This means that it
   might be possible to find a tiling pattern that has two hexagons and two triangles meeting at
   each point. See if you can find one using the shapes you cut out and draw a picture of it
   below.

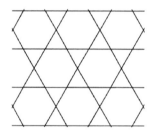

3. What addition problems correspond to the tiling patterns below? Can you think of
   interesting ways to color them in?

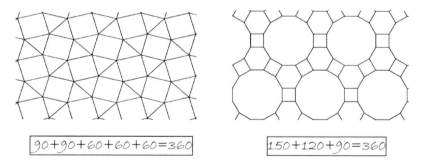

$90+90+60+60+60=360$ $150+120+90=360$

4. See if you can find a tiling pattern that corresponds to the addition problem 120 + 90 +90 +
   60 = 360 using the shapes you cut out.

   See question 6 on the next page for a picture.

5. Making Connections: Add each of the following in your head:

   $120 + 108 + 60 + 60 = \boxed{348}$

   $150 + 90 + 60 + 60 = \boxed{360}$

6. Challenge: The answer to question 4 on the previous page is that the regular tiling looks like the picture below. In the textbook I showed you pictures of two windows from the Blue Mosque. The two window patterns look different, but they are actually both obtained by coloring in this pattern. Can you see how? Can you also figure out how to color this to make it look like a pattern involving dodecagons, hexagons, and squares?

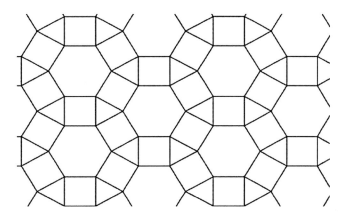

The window on the left is obtained by cutting a vertical stripe centered on a column of hexagons 4½ hexagons high from a larger tiling. The window on the right is obtained by turning the pattern on its side (counterclockwise.) To make a pattern with dodecagons, color a hexagon, the six surrounding squares, and the six surrounding triangles the same color. Then do the same with the hexagon directly to its right and to a hexagon above and in between.

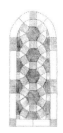
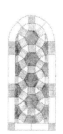

HMES

© 2013 Glenn Ellison

4.8 Other Tilings

1. The "home plate" on a baseball field is not a regular pentagon: it has three 90° angles and two 135° angles. Draw a picture to show how you could use a large number of home plates to tile a field.

2. The pentagon below has three 120° angles and two 90° angles. It is sometimes called the Cairo pentagon because there are streets in Cairo, Egypt that are paved with stones in this shape. Make some Cairo pentagons by tracing this shape and see if you can figure out how they tile.

3. One way to make up irregular shapes that will tile a floor is to start with a regular shape like a square or hexagon that tiles a floor. You then modify the shape by pushing out part of one side and pushing in part of the opposite side. For example, when I pushed out part of the top edge of the square below I did the same to the bottom edge. Try to design your own Escher-like fish tiling by modifying a brick to make it look more like a fish.

     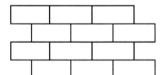

4. Challenge: What pattern do you get if you draw dots in the center of each tile in the tiling below and then connect each dot to the dots in the adjacent shapes?

Notes on 3: Try making a rounded triangular head sticking out on the right of each brick and make a similar cutout on the left to make a tail. Also squish the top and bottom in near the left and bulge them out on the right. On 4: The polygons made by connecting the dots are the "Cairo pentagons" from question 2. This also lets you see how Cairo pentagons can be used to tile the plane.

5.1 Look for Places Where There's Just One Possible Value

1. In each of the addition puzzles below there is at least one letter that has only one possible value. Say which letters they are and what value you would fill in.

```
     A              SEE              AAB
   + B            + BEE              AAC
   ───            ─────            + AAD
    CD             FLEE             ────
                                    684
```

$C = 1$ F=1, E=0   A=2

2. Replace each shape in the puzzle below with a different digit to make the addition correct:

```
    □ ☺ ☺ △            □ = 1
  + □ ☺ □ □            ☺ = 0
  ─────────            △ = 2
    2 0 1 3
```

3. Solve each of the number puzzles below.

```
     AB3                    A
   + 4A2                    A
   ─────                 + MA
    CCA                  ─────
                          SPA
```

A=5, B=4, C=9   S=1, M=9, P=0, A=5

4. A good first step for solving the puzzle below would be to notice that G = 1 is the only possibility for G. After this, you could realize that there must carrying in two different columns. Which columns are those? What can you then fill in in the thousands column?

```
     OLD          Columns that must have carrying: 100's, 1000's
   + BLUE         Values in the 000's column: B=_9_  R=_0_
   ─────
   GREEN
```

5. Making Connections: Solve the following base 8 addition puzzle:

```
    AA(8)
  +  B(8)            A=7, B=1, C=0
  ──────
   BCC(8)
```

Notes on 5: The connection is to section 2.6. To solve it, note first that from the left column you know B must be 1. Then, from the next column you know that A must be 7 (so there is carrying). Adding $77_{(8)} + 1_{(8)} = 100_{(8)}$ reveals that C must be 0.

5.2 Look for Places Where There Are Just A Few Possible Values

1. In the number puzzle below what values for the letter N are possible? _0 and 5_
    What values for M and C are possible? _1 and 2_ How do you know that there can't
    possibly be a 1 carried to the tens column? _1 + A + A would then be odd._

    Use these observations to help find a solution to the puzzle.

    ```
          MAN              M = 1
    +     CAN              A = 5
    ---------              N = 0
        4 0 0              C = 2
    ```

2. In the number puzzle below which letter will have to be an even number? _A_
    Will there be a 1 carried to the tens column? _Yes_ What values of E are possible? _Only 9._
    What values of F are possible? _Only 1_

    Using these observations find a solution to the puzzle.

    ```
        SEE        Answers for      S = 6    E = 9
    +   BEE        S, B, and L      F = 1    L = 4
    --------       can vary.        A = 8    B = 7
       FLEA
    ```

3. In each of the number puzzles below say which letters you can know things about even
    before you've solved the puzzle.

    ```
         HIT                    IS
    +    CAT              +    THIS
    --------             ---------
        MEAN                  GOOD
    ```

    Must be even: _N_   Must be even: _D_
    Must be 1: _M_   Must be 9: _H_
    Must be 0 or 9: _I_   Must be 0: _O_
                                   Must be at least 5: _I_

4. Making Connections: Replace each shape in the puzzle below with a different number to
    make the addition correct:

    ```
        □ ☺ △          □ = 1    Answers
        △ □ ☺          ◇ = 6    can vary.
    +   ☺ △ □          ☺ = 2
    ----------         △ = 3
      ◇ ◇ ◇
    ```

    Notes on 4: The connection is to section 1.4. All three columns are the same, so the puzzle
    will work if you choose any three different numbers that add up to less than 10 for □, ☺,
    and △: (1,2,3), (1,2,4), (1,2,5), (1,2,6), (1,3,4), (1,3,5), or (2,3,4) ordered in any way.

5.3 Be Organized While Trying Out Each Possible Value For Some Letter

1. In the puzzle below A must be either 8 or 9. (Can you see why?) Organize your work looking for a solution by trying A=8 on the left side of your scratch paper and A=9 on the right side. (A good second step is thinking about possible values of R.)

    ```
        ART          (L,T,R) can      A = 8    R = 3
      + FAT          also be (4,2,6)  T = 6    F = 1
       FILL          or (4,7,5)       I = 0    L = 2
    ```

2. In the puzzle below there are only two possible values for D: _0_ and _5_. Organize your work to find a solution by trying one possible value for D on the left side of your scratch paper and the other on the right side of your scratch paper.

    ```
        OLD          (G,O,L,D) can also be
        OLD          (3,7,5,0)
        OLD                            G = 1
        OLD                            O = 2
      + OLD                            L = 5
       GOLD                            D = 0
    ```

3. Find a solution to each of the number puzzles below.

    ```
        AAB   227        OLD      583       HIT      904
        AAC   228      + BLUE   + 9864    + CAT    + 754
      + AAD   229       GREEN    10447     MEAN     1658
        684   684                                Many solutions
      (or reorder B,C,D)                         possible.
    ```

4. Challenge: How many different solutions are there to the number subtraction puzzle below? Find ALL of them.

    ```
        ANNA       There are 4 solutions.    A = 1   N = 0
      -   TV       use V=4,5,6, or 7         T = 8   V =
        SAD        and D=11-V                S = 9   D =
    ```

Notes on 5: Think of the problem as an addition problem: SAD + TV = ANNA. First notice that A must be 1 and S must be 9. Then N must be 0 because ANNA is less than 999+99=1098, so the problem is 91D + TV = 1001. This means that TV + D = 91 which implies that T is 8 and V and D are two numbers that add up to 11. Neither can be 8 or 9 so the choices for (V, D) are (4, 7), (5, 6), (6, 5), and (7, 4).

5.4 Make Up Your Own Puzzles

1. Word Problem: Nelson wanted to make up a number puzzle using the words "BART" and "FART". He first randomly picked out the numbers 139 for "ART". He then wrote out the addition problem as I've done on the right below.

   He then realized that if he wanted the answer to have a vowel as the second letter, there was basically only one way to choose B and F: they had to be _4_ and _5_ in some order. When he filled in those and looked at the addition he realized that the first letter of the word problem answer would have to be _T_. Apart from this, he could choose any word for the answer as long as the last three letters weren't B, A, R, T, or F. What word would you have chosen?

   ```
     B  A  R  T      Answers          _  1  3  9
   + F  A  R  T      can vary.      + _  1  3  9
   ─  ─  ─  ─                         _  2  7  8
   ```

2. Make up two different number puzzles by choosing a different letter for each number in the addition problem below. First try one using A for the number 2 and then try one with some other letter there. (Don't forget that the two 7's and the two 3's have to be the same letter.)

   Many answers possible.

   ```
      727                  _  A  _                 _  _  _
   +  326      →       +   _  A  _     or      +   _  _  _
     1053                 ─  ─  ─  ─              ─  ─  ─  ─
   ```

3. Making Connections: One addition fact we learned when working with tilings was 150 + 120 + 90 = 360. Try to make a number puzzle based on it:

   ```
       90               _  _          Many answers possible.
      120               _  _  _       Try N, P, or T for 0.
   +  150      →     +  _  _  _
      360               _  _  _
   ```

4. Challenge: Some people have names that can be made into number puzzles: FIRSTNAME + MIDDLENAME = LASTNAME. Many others, unfortunately, cannot because their names are too different in length, or have more than 10 total letters, or have some other problem. Figure out whether you can make your name into a number puzzle. If not, try to think of someone else in your class who has a name that might work and make a name puzzle for them.

   Answers will vary.

Notes on 1: After picking 1,3,9 for ART the choices for B and F are 2,4,5,6,7,8. If 2 is chosen for either than the answer will have B or F as its second letter. So if you want 2 to be a vowel choosing 4 and 5 is the only way to make the sum at most 9. The answer can be any 4-letter word starting with T that does not use A, B, F or T (except at the start). On 2: Try picking something for 7 first to make a word, e.g. DAD, GAG. T and H are good choice for 3 because you can use words ending in ST or SH for the answer.

40

6.1 Memorizing the Multiplication Table: Start with the Really Easy Ones

1. Word Problem: All of the Penderwick children are girls. The number of Penderwick daughters is two more than the number of Geiger sons. The number of Geiger sons is two more than the number of Penderwick sons. How many toes do the children in the Penderwick family have all together?

   $\boxed{40}$

2. Find the answers to each of the multiplication problems below. Try to do them as quickly as you can and time yourself to see how long it takes you.

5×10	3×10	1×10	10×7	4×10	10×9
50	30	10	70	40	90

2×10	8×10	6×10	10×4	10×10	7×10
20	80	60	40	100	70

3×10	9×10	5×10	10×1	8×10	Time:_____
30	90	50	10	80	

3. Get someone to quiz you on × 10's using flashcards or practice on your own using the website http://arithmetic.zetamac.com/ or some other website. (On the zetamac site change the check boxes so only multiplication is checked and change the settings so the first number can be anything from 1 to 10 and the second is always 10.) See how many problems you can answer correctly in 120 seconds and record the total here. Then try a couple more times to see if you can improve your record.

Date	# of ×10's in 120 seconds

4. Making Connections: Find 30 + 31 + 32 + 33 + 34 + 35 + 36 + 37 + 38 + 39.

   $\boxed{345}$

Notes on 1: The first sentence means that there are zero Penderwick boys. Then work backwards. On 4: One way to do this is to first connect to section 3.6 – add the tens then the ones. The digits in the tens place are 10 3's so they add up to 30 tens, which is 300. Then add 0+1+2+3+4+5+6+7+8+9 as in section 3.3 – adding evenly spaced numbers.

     © 2013 Glenn Ellison

6.2 Learn Your ×2's and ×3's By Skip Counting

1. Word Problem: Amy, Allison, and Anying go to the mall. Each buys seven pairs of socks. How many pairs of socks did they buy all together?

21

2. Find the answers to each of the multiplication problems below. Try to do them as quickly as you can and time yourself to see how long it takes you.

5	3	1	2	4	2
× 2	× 2	× 2	× 7	× 2	× 9
10	6	2	14	8	18

2	8	6	2	2	7
× 2	× 2	× 2	× 4	× 10	× 2
4	16	12	8	20	14

2	2	5	2	8	
× 10	× 2	× 2	× 1	× 2	Time:_____
20	4	10	2	16	

3. Get someone to quiz you on ×2's and ×3's using flashcards or practice on your own using http://arithmetic.zetamac.com/ or some other website. See how many you can answer in 120 seconds. Then keep trying to beat your record. You won't get as many as you got doing 10's, but see if you can come close.

Date	# of ×2's	Date	# of ×3's	Date	Mixed 2's & 3's

HMES   © 2013 Glenn Ellison

6.2 Learn Your ×2's and ×3's By Skip Counting

4. Find the answers to each of the multiplication problems below. Try to do them as quickly as you can and time yourself to see how long it takes you.

$$
\begin{array}{r} 5 \\ \times\ 3 \\ \hline 15 \end{array}
\qquad
\begin{array}{r} 3 \\ \times\ 3 \\ \hline 9 \end{array}
\qquad
\begin{array}{r} 1 \\ \times\ 3 \\ \hline 3 \end{array}
\qquad
\begin{array}{r} 3 \\ \times\ 7 \\ \hline 21 \end{array}
\qquad
\begin{array}{r} 4 \\ \times\ 3 \\ \hline 12 \end{array}
\qquad
\begin{array}{r} 3 \\ \times\ 9 \\ \hline 27 \end{array}
$$

$$
\begin{array}{r} 3 \\ \times\ 2 \\ \hline 6 \end{array}
\qquad
\begin{array}{r} 8 \\ \times\ 3 \\ \hline 24 \end{array}
\qquad
\begin{array}{r} 6 \\ \times\ 3 \\ \hline 18 \end{array}
\qquad
\begin{array}{r} 3 \\ \times\ 4 \\ \hline 12 \end{array}
\qquad
\begin{array}{r} 3 \\ \times\ 10 \\ \hline 30 \end{array}
\qquad
\begin{array}{r} 7 \\ \times\ 3 \\ \hline 21 \end{array}
$$

$$
\begin{array}{r} 3 \\ \times\ 10 \\ \hline 30 \end{array}
\qquad
\begin{array}{r} 3 \\ \times\ 2 \\ \hline 6 \end{array}
\qquad
\begin{array}{r} 5 \\ \times\ 3 \\ \hline 15 \end{array}
\qquad
\begin{array}{r} 3 \\ \times\ 1 \\ \hline 3 \end{array}
\qquad
\begin{array}{r} 8 \\ \times\ 3 \\ \hline 24 \end{array}
\qquad
\text{Time:}\_\_\_\_\_
$$

5. Making Connections: What is $30_{(8)}$ in base 10?

24

6. Challenge: Starting at 30 count backwards by 3's until you get to the smallest positive number you'll get to. Then count up from there by 2's until you get to the largest number less than or equal to 30 that you'll get to. Then count backwards from there by 3's until you get to the smallest positive number you'll get to. Keep going back and forth until you get to say 30 again. Then stop. Oh, I forgot to say that you should count on your fingers the whole time to keep track of how many numbers you are saying. How many numbers did you say in total (including the 30 at the start and the end)?

46

Notes on 5: The connection is to section 2.7. Recall that $30_{(8)}$ just means 3 eights and 0 ones, so to convert to base 10 you just multiply 3×8. On 6: The first 10 numbers are 30, 27, …, 3. Then counting up 5, 7, …, 29 is 13 more numbers. Going down 26, 23, …, 2 is 9 more numbers. Then counting up 4, …, 30 is 14 more. The total number of numbers is $10 + 13 + 9 + 14 = 46$.

6.3 Rows with Patterns: Learning the ×5's and ×9's

1. Word Problem: Yu Shen wants to buy a package of chocolate chip cookies which costs $1.29 He has 9 dimes and 8 nickels in his pocket. Is this enough money?

    > Yes. He has $1.30.

2. Find the answers to each of the multiplication problems below. Try to do them as quickly as you can and time yourself to see how long it takes you.

$$
\begin{array}{cccccc}
5 & 3 & 5 & 5 & 5 & 5 \\
\times\ 5 & \times\ 5 & \times\ 5 & \times\ 5 & \times\ 9 & \times\ 9 \\
\hline
25 & 15 & 25 & 25 & 45 & 45
\end{array}
$$

$$
\begin{array}{cccccc}
5 & 8 & 5 & 5 & 5 & 8 \\
\times\ 2 & \times\ 5 & \times\ 2 & \times\ 8 & \times\ 10 & \times\ 5 \\
\hline
10 & 40 & 10 & 40 & 50 & 40
\end{array}
$$

$$
\begin{array}{ccccc}
5 & 5 & 5 & 7 & 4 \\
\times\ 10 & \times\ 7 & \times\ 6 & \times\ 5 & \times\ 5 \quad \text{Time:}\_\_\_\_ \\
\hline
50 & 35 & 30 & 35 & 20
\end{array}
$$

3. Get someone to quiz you on ×5's and ×9's using flashcards or practice on your own using http://arithmetic.zetamac.com/ or some other website. See how many you can answer in 120 seconds. Then keep trying to beat your record. See if you can do as well as you did with the 2's and 3's.

Date	# of ×5's	Date	# of ×9's

Notes on 1: The value of the coins in cents is $(9 \times 10) + (8 \times 5) = 90 + 40 = 130$.

6.3 Rows with Patterns: Learning the ×5's and ×9's

4. Find the answers to each of the multiplication problems below. Try to do them as quickly as you can and time yourself to see how long it takes you.

5	3	5	6	9	9
× 9	× 9	× 9	× 9	× 9	× 9
45	27	45	54	81	81

9	9	9	9	9	8
× 2	× 4	× 2	× 8	× 10	× 9
18	36	18	72	90	72

4	9	9	7	6	
× 9	× 7	× 1	× 9	× 9	Time: _____
36	63	9	63	54	

5. Making Connections: Find the sum of all two digit numbers that have a 9 in them or are a multiple of 9:

    $18 + 19 + 27 + 29 + \ldots + 99$

    $\boxed{1773}$

6. Challenge: If you start counting by 9's the first several numbers, 9, 18, 27, …, all have the sum of their digits equal to 9. The next number 99 doesn't. But after this the next several 108, 117, 126, … again have digits that add up to 9. What is the fourth number in this sequence that has the sum of its digits not equal to 9?

    $\boxed{279}$

Notes on 5: Making connections to sections 1.4 and 3.3 helps, but this is still a hard problem. Start by separating the numbers to be added into three nonoverlapping sets: numbers starting with 9; numbers less than 99 ending with 9; and multiples of 9 less than 90. $(90 + 91 + \ldots + 99) + (19 + \ldots + 89) + (18 + 27 + 36 + \ldots + 81)$. The first is 945 (add the tens then the ones). The second is 360+72=432 (ditto). The third is $360 + 36 = 396$ (as in section 1.4 there are 36 ones and then 36 tens). On 6: The main idea was just to have students practice counting by 9's. After 99, the next multiples whose digits don't add up to 9 are 189, 198, and then 279.

6.4 Finishing Off the Table: Multiplying by 4, 6, 7, and 8

1. Word Problem: The Bigelow math team sends 6 girls and 4 boys to an IMLEM math meet. Each girl scores 8 points on the individual rounds. Each boy scores 6 points on the individual rounds. And then the team gets 5 questions right on the team round (where questions are worth 6 points each). What is the team's total score?

    $\boxed{102}$

2. Find the answers to each of the multiplication problems below. Try to do them as quickly as you can and time yourself to see how long it takes you.

4 × 4 = 16	4 × 6 = 24	4 × 4 = 16	4 × 6 = 24	4 × 4 = 16	4 × 6 = 24
4 × 6 = 24	4 × 7 = 28	4 × 6 = 24	4 × 7 = 28	4 × 5 = 20	4 × 7 = 28
4 × 8 = 32	4 × 4 = 16	4 × 8 = 32	4 × 9 = 36	8 × 4 = 32	Time:____

3. Get someone to quiz you on ×4's, ×6's, ×7's, and ×8's using flashcards or practice on your own using http://arithmetic.zetamac.com/ or some other website. See how many you can answer in 120 seconds. Then keep trying to beat your record. Do some extra practice on whichever one you're worst at.

Date	# of 4's	Date	# of 6's	Date	# of 7's	Date	# of 8's

Notes on 1: The total score was $(8 \times 6) + (4 \times 6) + (5 \times 6) = 48 + 24 + 30 = 102$.

6.4 Finishing Off the Table: Multiplying by 4, 6, 7, and 8

4. Find the answers to each of the multiplication problems below. Try to do them as quickly as you can and time yourself to see how long it takes you.

6 × 6 **36**	6 × 7 **42**	6 × 6 **36**	6 × 7 **42**	6 × 6 **36**	6 × 7 **42**
6 × 8 **48**	7 × 8 **56**	8 × 6 **48**	7 × 8 **56**	6 × 8 **48**	7 × 8 **56**
7 × 7 **49**	8 × 8 **64**	7 × 7 **49**	8 × 8 **64**	6 × 9 **54** Time:_____	

5. Making Connections: Grace cuts out 4 squares, 6 hexagons, and 8 octagons. If she makes one cut along each side of each shape, how many cuts does she make?

116

6. How many numbers in the 50's are the answer to at least one ×4, ×6, ×7, or ×8 multiplication problem?

7. Challenge: Noam was bored in school one day so he started adding together all of the numbers in a 10 × 10 multiplication table. First, he added the numbers in the ×1 row, then he added all the numbers in the ×2 row, and so on. After a while he noticed a pattern that made things easier and he finished the whole table before class was over. What answer did he get?

3025

Notes on 5: One connection is remembering how many sides each shape has. The number of cuts is (4 × 4) + (6 × 6) + (8 × 8) = 16 + 36 + 64. It also connects to section 3.2: it's easier if you add this as 16 + (36 + 64) = 16 + 100 = 116. On 7: The sums of the numbers in each row are 55, 110, 165, 220, 275, 330, 385, 440, 495, and 550. (The patterns are more obvious if you look at every other number: 110, 220, 330, 440, 550; and 55, 165, 275, 385, 495.) One way to add them is with addition with carrying. In the ones column there are five 5's and five 0's which add up to 25. So put a 5 and carry the 2. In the tens column there numbers are 1+2+3+4+5+6+7+8+9+5 = 50, so put a 2 and carry the 5. In the hundreds column the numbers are 1+1+2+2+3+3+4+4+5 = 25, so put a 0 and carry the 3.

       © 2013 Glenn Ellison

7.1 Prime and Composite Numbers

1. For each of the numbers below write down all multiplication problems that have the number as the answer. Then say if there are fewer than two, exactly two, or more than two. Then say whether the number is prime, composite, or neither

Number	Multiplication problems with this answer	Less than two, Exactly two, or More than two	Prime, Composite, or Neither
3	1×3 and 3×1	Exactly 2	Prime
4	1×4, 2×2, and 4×1	More than 2	Composite
6	1×6, 2×3, 3×2, and 6×1	More than 2	Composite
7	1×7 and 7×1	Exactly 2	Prime
15	1×15, 3×5, 5×3, and 15×1	More than 2	Composite
24	1×24, 2×12, 3×8, 4×6, 6×4, 8×3, 12×2, and 24×1	More than 2	Composite
1	1×1	Less than 2	Neither

2. Name a number that is not prime even though you will find it exactly twice in a 10 × 10 multiplication table.

Answers can vary. Some examples are 14, 15, 21, 27, 28, …

3. There isn't enough material in section 7.1 for me to ask a whole worksheet worth of questions. But your teacher may not know this if she hasn't read the chapter or this worksheet. So just play a game of hangman before doing the final problem.

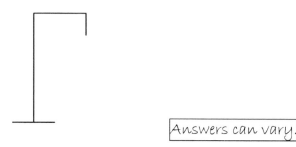

Answers can vary.

4. Making Connections: How do you know that 360 is not prime?

In Chapter 4 we saw examples of 360 being divided evenly. For example, 360 = 6 ×60, 360 = 4 × 90, and 360 = 3 × 120.

7.2 Figuring Out What's Prime: Just Look in the Multiplication Table

1. Word Problem: Arya recently got 19 zoots. She wants to hold a party at which she will invite over at least one friend and she will share out the zoots so that she and all her friends each get the same number of zoots and they all get more than one. Is this possible?

No

2. Determine whether the following numbers are prime or composite. Circle the prime numbers. Put a big X through each composite number and write a multiplication problem below it that proves it is composite.

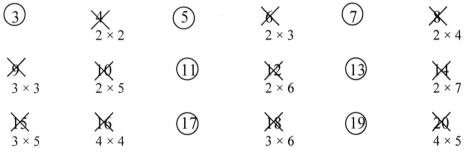

③   ✗   ⑤   ✗   ⑦   ✗
          2×2 2×3 2×4

✗   ✗   ⑪   ✗   ⑬   ✗
3×3 2×5 2×6 2×7

✗   ✗   ⑰   ✗   ⑲   ✗
3×5 4×4 3×6 4×5

(Some listed multiplications can vary.)

3. Making Connections: Is $1 + 2 + 3 + 4 + 5 + 6 + 7 + 8 + 9$ prime?

No

4. Challenge: What is the smallest number that is composite even though you won't find it as an answer in 10×10 multiplication table?

22

Notes on 3: The connection is to section 3.3. The numbers add up to 45 which is not prime. A more general result is that the sum of three or more consecutive positive whole numbers is never prime. If you add an odd number of numbers the sum is equal to the number of numbers times the middle number. If you add an odd number of numbers it's a multiple of the sum of the first and last number. On 4: If a number is composite and does not appear in a 10 by 10 multiplication table then it is a product of two numbers each of which is at least two and one of which bigger than 10. Clearly 2×11 would be the smallest number that meets these requirements. It is indeed not found in a 10 by 10 multiplication table so it is the desired number.

7.3 Figuring Out What's Prime: Think About What the Factors Could Be

1. Word Problem: Cersei is trying to guess a number Jaime is thinking of. Jaime says his number has is odd, not prime, and is a multiple of two different prime numbers. What is the smallest possible number Jaime could be thinking of?

2. Determine whether the following numbers are prime or composite. Circle the primes. Put a big X through each composite and write a multiplication problem that proves it is composite.

    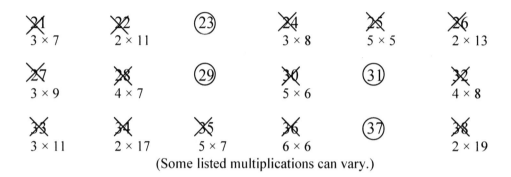

    (Some listed multiplications can vary.)

3. Is 11 prime? How about 33? How about 555? How about 7777?

    Yes. No. No. No.

4. Making Connections: Which of the following base 8 numbers are prime and which are not? Circle the primes.

    $10_{(8)}$ $11_{(8)}$ (15$_{(8)}$)   (23$_{(8)}$) $25_{(8)}$

5. Challenge: What is the smallest composite number that is not a multiple of 2, 3, 5, or 7?

    Notes on 3: The last three are not prime because 33 = 3 × 11, 555 = 5 × 111 = 5 × 3 × 37, and 7777 = 7 × 1111 = 7 × 11 × 101. On 4: The connection is to section 2.7. In base 10 the numbers are 8, 9, 13, 19, and 21. On 5: If a number is composite and not a multiple of 2, 3, 5, and 7, then it must have two prime factors each of which is at least 11. The only tricky part is that you need to realize that you can use two 11's so the answer is 11 × 11 = 121, rather than 11 × 13 = 143.

7.4 What Can I Do When I Know About Prime Numbers?

1. Word Problem: Aly has a garden with a fence on 3 sides of it. There are 11 feet of fencing and her garden is 14 square feet. What is the length of the shorter side of her garden?

   | 2 feet |

2. Cedric has rectangular handkerchief that is 21 square inches and has 20 inches of border. What is the length of the shorter side of the rectangle?

   | 3 inches |

3. The Weasley Family has 9 members. If they want to divide up into multiple teams of more than one player each to play a game, can they do this? What if the two parents don't play?

   | Yes (3 teams of 3). No if the parents don't play. |

4. Making Connections: Many times when we measure angles, we say that 360 degrees is all the way around a circle. This was decided because 360 has many factors, and so you can divide it into 2, 3, 4, 5, 6, 8, 9, 10, and many other numbers of angles with the angle still having an integer degree measure. If one were to create a new system of measuring angles, which of the following numbers of degrees would be best? Which would be worst? Why?

   111   120   122   124   132   137

   | 120 is probably best – it can be divided by 2, 3, 5, 6, 8, 10, etc. |

   | 137 is probably worst. It is prime. |

5. Challenge: Molly has a rectangular garden with an area of 175 square feet. Both the length and the width (in feet) are whole numbers. She wants to put a fence along all four sides of the garden. What are all the possibilities for the length of fencing she might need?

   | 64 feet (7 × 25), 80 feet (5 × 35), 352 feet (1 × 175) |

Notes on 1: The only ways to get an area of 14 square feet with integer side lengths is with a 2 by 7 or 1 by 14 garden. Here, 2 by 7 works – the fencing is along both short sides and one long side. On 2: The rectangle is 3 inches by 7 inches. On 5: It helps to think about the prime factorization of $175 = 5 \times 5 \times 7$. The shorter side will need to be 1, 5 or 7 feet. The longer side will need to be a product of the other two numbers (or all three if the short side is one foot.)

7.5 Which Numbers Are Prime?

1. Word problem: Ella is having a birthday party. She can invite as few as 10 people or as many as 50. However, all of her friends like to play a game which requires that there be multiple teams of at least 2 people. How many numbers of people can she invite?

    30

2. How many 1 or 2 digit primes end in a 0? _____0_____ a 1?___5___ a 2?___1___ a 3?___7___
    a 4?_____0_____ a 5?___1___ a 6?___0___ a 7?___6___ an 8? _____0_____ a 9?___5___

3. What is the smallest prime bigger than 17?

    19

4. How many primes are between 40 and 50?

    3

5. The smallest prime number p for which p+10 and p+20 are also prime is 3. What is the smallest prime number p for which p + 30 and p + 60 are also prime?

    7

6. Making Connections: What prime is closest to 367 − 242 + 33 − 58 + 4 − 10?

    97

7. Challenge: The first five primes – 2, 3, 5, 7, and 11 – are all palindromes: they stay the same if you reverse their digits. After this, there are N primes that are not palindromes before you get to the next prime palindrome. What is N?

    20

    _____
    Notes on 1: She can pick any composite number from 10 to 50. There are 41 numbers from 10 to 50 and 11 of them are prime: 11, 13, 17, 19, 23, 29, 31, 37, 41, 43, 47. On 6: The connection is to section 3.2. The additions and subtractions are easier if you reorder as (367+33) − (242+58) + 4 − 10 = 400 − 300 + 4 − 10 = 94. On 7: All two digit palindromes after 11 are not prime (because 22, 33, … are all multiples of 11). The first three digit palindrome, 101, is prime. So the answer to the questions is just the number of primes between 11 and 101. There are 20 of them.

7.6 Twin Primes

1. Word problem: Em is exactly two years younger than Henry. If Em is 42 when she and Henry go to Oz and stop aging, how many times will their ages have both been prime numbers at the same time?

2. List all the pairs of twin primes less than 100.

(3,5), (5,7), (11,13), (17,19), (29,31), (41,43), (59,61), (71,73)

3. Cousin primes are primes that differ by 4. How many pairs of cousin primes are there where both numbers are less than 100?

8

(They are (3,7), (7,11), (13,17), (19,23), (37,41), (43,47), (67,71), and (79,83).)

4. Challenge: What is the largest pair of twin primes smaller than 300? What is the smallest pair of twin primes larger than 300?

(281, 283) and (311, 313)

Notes on 1: The question is asking for the number of pairs of twin primes with the smaller number at most 42. There are 6 such pairs: (3, 5), (5, 7), (11, 13), (17, 19), (29, 31), and (41, 43). On 4: A good way to work on these problems is to think about the pattern of where in every 30 numbers you might find twin primes: the numbers that are (11, 13), (17, 19), and (29, 31) more than a multiple of 30. 270 is a multiple of 30 so the three pairs of numbers that might be twin primes in the 280's, 290's, and 300's are (281, 283), (287, 289), and (299, 301). 289 is not prime (it's 17 × 17) so the largest pair less than 300 is (281, 283). To look for a pair greater than 300 note that in the 310's, 320's, and 330's the possible twin primes are (311, 313), (317, 319), and (329, 331). The first pair does work.

8.1 Subtraction

1. Word Problem: If Ender led 94 ships into battle against aliens, and 82 were destroyed, how many ships were not destroyed?

   $\boxed{12}$

2. Rewrite the following problems with one number on top of the other and subtract.

   $234 - 123 = 111$ $563 - 421 = \boxed{142}$ $829 - 17 = \boxed{812}$

   234
   -123
   $\overline{111}$

   $777 - 673 = \boxed{104}$ $8243 - 31 = \boxed{8212}$ $9439 - 1029 = \boxed{8410}$

3. Rewrite the following problems with one number on top of the other and subtract. Don't forget to line things up on the decimal points.

   $124.23 - 12.21 = 112.02$ $894142 - 71031 = \boxed{823{,}111}$

   124.23
   -12.21
   $\overline{112.02}$

   $938.32 - 834.2 = \boxed{104.12}$ $5589675.974 - 172445.22 =$

   $\boxed{5{,}417{,}230.754}$

4. Making Connections: An interior angle of a regular dodecagon measures 150°. An interior angle of a regular decagon is 144°. An interior angle of a regular hexagon is 120°. Which is larger: the angle that you still need to fill in to make 360° after you put two decagons together or the angle you still need to fill in when you put a hexagon next to a dodecagon? How much bigger is it?

   $\boxed{\text{The angle to fill in next to the hexagon and dodecagon is 18° bigger.}}$

   Notes on 4: The connection is to sections 4.4 and 4.5 (but nothing from those sections is needed to do the problem). The question asks which is bigger $360 - (150 + 120)$ or $360 - (144 + 144)$. You can find this by just doing the calculations. But it's quicker if you just realize that $150 + 120$ is smaller than $144 + 144$ so $360 - (150 + 120)$ is larger.

8.2 Subtraction with Borrowing

1. Word Problem: In the Olympics, Shawn Johnson did a floor routine with a maximum score of 16.3, but she made a big mistake and the judges deducted 1.4 from her maximum score. What was her final score?

$$\boxed{14.9}$$

2. Subtract the following.

$2954 - 1993 = 961$ $271 - 139 = \boxed{132}$ $452 - 271 = \boxed{181}$

```
 1 18 15
 2954
-1993
  961
```

$5621 - 39 = \boxed{5582}$ $56554 - 5655 = \boxed{50,899}$ $2101 - 1872 = \boxed{229}$

$10002 - 9384 = \boxed{618}$ $4510044 - 3625931 = \boxed{884,113}$

3. Making Connections: Subtract the sum of all positive multiples of 7 less than 100 from the sum of all positive multiples of 5 less than 100.

$$\boxed{215}$$

4. Challenge: On November 10, 2084 the Earth will pass directly between Mars and the Sun. If there are people on Mars (or any Martians there) they will see the Earth and the Moon as black dots blocking out a small part of the Sun. On that date the Earth will be 148,153,060 kilometers from the Sun. Mars will be something like 227,900,000 kilometers away from the Sun. If Mars were that far away from the Sun on that date, how far would Earth be from Mars?

$$\boxed{79,746,940 \text{ kilometers}}$$

Notes on 4: The connection is to section 3.3. The sums $7 + 14 + \ldots + 98$ and $5 + 10 + \ldots + 95$ can both be computed by adding pairs of numbers from opposite ends. The first is $105 \times 7 = 735$ and second is $9 \times 100 + 50 = 950$. Then $950 - 735 = 215$. On 5: When the Earth is directly between Mars and the Sun you can find the Mars to Earth distance just by subtracting $227,900,000 - 148,153,060$. Doing the subtraction is a good way to see if students understand how to do borrowing when you have to borrow in multiple columns in a row.

8.3 Subtracting in Your Head

1. Word Problem: Wanda got 249 pounds of flour on one particular excursion. If she and her friends then used 124 pounds, how much flour was left?

   125 pounds

2. Suppose you are doing the subtraction problem 671 – 253 in your head by first subtracting the 200, then subtracting the 50, then subtracting 3. To help keep track of things, you say the answer to yourself after each subtraction. Write down the list of numbers you say to yourself in order.

   471, 421, 418

3. Subtract the following in your head.

   295 – 126 = 169   844 – 411 = 433 452 – 248 = 204

   512 – 256 = 256   565 – 393 = 172 2754 – 1522 = 1232

   3552 – 1334 = 2218   45,400 – 33,259 = 12,141

4. Making Connections: Arjun puts one edge of a regular decagon (which has a 144° angle) up against the edge of a regular pentagon (which has a 108° angle) with the same edge-length. Is there a regular polygon that would exactly fit the remaining space where the angles come together?

   Yes. A regular pentagon will fit.

5. Challenge: Eunice made a big number by typing the digits in the order they appear along the top of a typewriter keyboard: 1234567890. She then made another big number by writing the numbers 13 through 17 next to each other with no spaces. And then she subtracted the smaller number from the larger one. Figure out the answer she got in your head.

   79,583,727

Notes on 4: The connection is to sections 4.4 and 4.5. The two angles add up to 252 degrees. 360 – 252 = 108. This is the angle in a regular pentagon. On 5: The question asks you to compute 1314151617 – 1234567890. There's no easy way to do this. One approach is to use subtraction with borrowing and go right to left. Another is to go one digit at a time from left to right saying the number at each step: 1314151617, 314151617, 114151617, ... The way I find easiest is to break it up after 4 digits because of the similarity in how the numbers start: 1,314,000,000 – 1,234,000,000 + 151617 – 567890. The first term is 80,000,000. But the second one is negative so I think of the answer as 79,000,000 + 151,617 + 1,000,000 – 567,890 = 79,0000,000 + 151,617 + 432,110 = 79,583,727.

8.4 More Shortcuts: The 00 and 999 methods

1. Word Problem: On a train ride, Harry buys 624 pieces of candy but gives 478 of them to Ron. How many pieces does he keep?

   146

2. Figure out the answer to each of the following subtraction using the 00 method: figure out how much bigger than 500 the bigger number is, figure out what you need to add to the smaller number to make 500, and then adding the two numbers together.

   $723 - 468 = 255$ $612 - 497 = 115$ $873 - 482 = 391$

   $723 - 500 = 223$ $612 - 500 = 112$ $873 - 500 = 373$

   $468 + 32 = 500$ $497 + 3 = 500$ $482 + 18 = 500$

   $223 + 32 = 255$ $112 + 3 = 115$ $373 + 18 = 391$

3. Subtract following in your head using the 00 method:

   $314 - 188 = 126$ $532 - 191 = 341$ $2341 - 279 = 2062$

   $4456 - 1398 = 3058$ $5717 - 4185 = 1532$

4. Write out the subtractions that you'd use to solve the problem below using the 999 method and find the answer.

   $3358 - 1779 = (3358 - 3000) + (3000 - 2999) + (2999 - 1779) = 358 + 1 + 1220 = 1579$

   $855 - 478 = (855 - 500) + (500 - 499) + (499 - 478) = 355 + 1 + 21 = 377$

   $3456 - 1888 = (3456 - 2000) + (2000 - 1999) + (1999 - 1888) = 1456 + 1 + 111 = 1568$

5. Making Connections: Arjun added together the smallest pair of twin primes greater than 50. He then subtracted the largest prime number that is less than 100. What answer did he get?

   23

Notes on 5: The connection is to section 7.6. To find the smallest twin primes greater than 50 a good approach is to remember that the three places where you might find twin primes are (11, 13), (17, 19) and (29, 31) after each multiple of 30. Adding 30 to each of these the first possibility after 50 is (59, 61) which are both primes, so this is a pair of twin primes. The largest prime less than 100 is 97. So the answer is $59 + 61 - 97 = 120 - 97 = 23$.

8.5 Reordering and Telescoping Sums

1. Word Problem: Ciara has 387 pennies, 149 nickels, and 67 dimes. Lucia has 145 pennies, 65 nickels, and 286 dimes. How many more coins does Ciara have than Lucia?

    107

2. Find the following.

   $(1 + 3 + 5 + \ldots + 21) - (2 + 4 + 6 + \ldots + 20) = 11$

   $(292 + 493 + 106) - (105 + 282 + 492) = \boxed{12}$

   $(834 + 219 + 1002 + 106) - (119 + 802 + 833) = \boxed{407}$

   $(3 + 5 + 7 + 9 + \ldots + 31) - (1 + 3 + 5 + 7 + \ldots + 29) = \boxed{30}$

   $(17 + 27 + 37 + \ldots + 77) - (13 + 23 + 33 + \ldots + 73) = \boxed{28}$

3. Find the sum of the differences between each pair of adjacent primes from 2 to 37, i.e. find $(3 - 2) + (5 - 3) + \ldots + (37 - 31)$.

   $\boxed{35}$

4. What is the difference between the sum of the first seven even multiples of 7 and the sum of the first seven odd multiples of 7?

   $\boxed{49}$

5. Making Connections: Find the difference between the sum of the multiples of 5 from 1 to 100 and the sum of the multiples of 4 from 1 to 80.

   $\boxed{210}$

6. Challenge: Find $\dfrac{1}{2} + \dfrac{1}{4} + \dfrac{1}{8} + \dfrac{1}{16} + \dfrac{1}{32} + \dfrac{1}{64} + \dfrac{1}{128} + \dfrac{1}{256} + \dfrac{1}{512}$.

   $\boxed{511/512}$

Notes on 1: A good reordering is $(387 - 286) + (149 - 145) + (67 - 65)$. On 3: If you write the subtractions in the opposite order, $(37 - 31) + (31 - 29) + (29 - 23) + \ldots + (3 - 2)$, it's more obvious that it telescopes and you're just left with $37 - 2 = 35$. On 4: Each even multiple is 7 bigger than the odd multiple, so the sum is $7 \times 7 = 49$. On 5: The connection is to section 3.3. You could compute both sums separately, but an even easier way is to think of the sum as $(5 - 4) + (10 - 8) + (15 - 12) + \ldots + (100 - 80) = 1 + 2 + 3 + \ldots + 20$. Now adding by the method of section 3.3 gives $21 \times 10 = 210$. On 6: We really haven't covered fractions, so the thing to do is to notice the pattern of the sums: 1/2, 3/4, 7/8, …

8.6 Borrowing in Base 8

1. Word Problem: Lisa Simpson printed out $73_{(8)}$ crossword puzzles and then did $65_{(8)}$ of them. How many did she have left to do?

2. Subtract the following (All numbers in this section are in base 8, even if they don't have an $_{(8)}$ after them.)

   $321_{(8)} - 206_{(8)} = 113_{(8)}$ $54_{(8)} - 42_{(8)} = \boxed{12_{(8)}}$

   $\begin{array}{r} 1\ 9 \\ 3\cancel{2}\cancel{1}_{(8)} \\ -\ 206_{(8)} \\ \hline 113_{(8)} \end{array}$

   $30_{(8)} - 24_{(8)} = \boxed{4_{(8)}}$ $72_{(8)} - 16_{(8)} = \boxed{54_{(8)}}$

   $552_{(8)} - 225_{(8)} = \boxed{325_{(8)}}$ $620_{(8)} - 465_{(8)} = \boxed{133_{(8)}}$

   $1000_{(8)} - 563_{(8)} = \boxed{215_{(8)}}$ $7405_{(8)} - 6666_{(8)} = \boxed{517_{(8)}}$

3. Making Connections: Use a base 8 version of the 00 method for mental subtraction to figure out $734_{(8)} - 475_{(8)}$ in your head.

   $\boxed{237_{(8)}}$

4. Challenge: Replace each letter with a different digit to make the following base 8 subtraction problem true.

   $\begin{array}{r} \text{NINE}_{(8)} \\ -\ \ \text{TEN}_{(8)} \\ \hline \text{TWO}_{(8)} \end{array}$

N	= 1	I	= 3
E	= 7	T	= 5
W	= 2	O	= 6

   Notes on 3: The connections are to sections 2.6 and 8.4. $475_{(8)}$ is 3 less than $500_{(8)}$. So to use the base 8 version of the 00 method you compute the subtraction as $734_{(8)} - 500_{(8)} + 3_{(8)}$ $= 237_{(8)}$. On 4: First observe that N must be 1. Next, note there must be borrowing from the 512's and 64's columns: there is borrowing from the N to make it 0; and in the last two columns 1E must be smaller than EN because if E=0 then W is 0 or 1. Using I + 8 – 1 – T = T the only possible values of (T, I) are (5, 3) and (6, 5). The first pair turns out to work. You know E is one more than O so the only possible choices for O are 6 and 7. 6 turns out to work.

9.1 The Distributive Property

1. Word Problem: In the first half of the Patriots game against Denver the Patriots scored 7 touchdowns, each worth 7 points. Suppose that they had scored 6 more touchdowns in the second half. How many points would they have scored in the game if this happened?

    $\boxed{91}$

2. Write out how you would compute the products below using the distributive property (but don't actually find the answers).

    $22 \times 7 = (20 \times 7) + (2 \times 7)$ $31 \times 8 = \boxed{(30 \times 8) + (1 \times 8)}$

    $103 \times 4 = \boxed{(100 \times 4) + (3 \times 4)}$ $121 \times 6 = \boxed{(100 \times 6) + (20 \times 6) + (1 \times 6)}$

3. Using the distributive property, find the following.

    $12 \times 4 = 48$ $11 \times 7 = \boxed{77}$

    $13 \times 8 = \boxed{104}$ $11 \times 13 = \boxed{143}$

    $41 \times 4 = \boxed{164}$ $101 \times 7 = \boxed{707}$

4. Using the distributive property with subtraction find the following.

    $19 \times 5 = \boxed{95}$ $98 \times 8 = \boxed{784}$

    $971 \times 6 = \boxed{5826}$

5. Making Connections: What is $(288 + 208 + 312) \times 6$?

    $\boxed{4848}$

6. Challenge: Use the distributive property to find the first 7 powers of 11. What patterns do you see in the answers?

    $\boxed{11, 121, 1331, 14641, 161051, 1771561, 19487171}$

    Notes on on 4: The last multiplication is easier if you use the distributive property to say $971 \times 6 = (1000 \times 6) - (30 \times 6) + (1 \times 6)$. On 5: The connection is to section 3.2. The addition is easier if you reorder and add $288 + 312 + 208$. On 6: Compute all the powers in order. I like to use addition with carrying to add numbers like $13310 + 1331$. Some patterns are: all numbers start and end with 1; the sum of the digits in the even columns are same as the sum of the digits in the odd columns, e.g. in 14641 we have $1 + 6 + 1 = 4 + 4$. Some patterns that are true for the first few but then don't continue are that the numbers are palindromes and the sum of the digits of the nth number is 2^n.

9.2 The Two- Sided Distributive Property

1. Word Problem: If a package of Skittles costs 91 cents, how much would it cost to buy 12 packages?

    $\boxed{\$10.92}$

2. Write how you would compute the products using the distributive property.

    $22 \times 13 = (20+2) \times (10+3) = 20\times10 + 2\times10 + 20\times3 + 2\times3$

    $52 \times 21 = (50+2) \times (20+1) = 50\times20 + 2\times20 + 50\times1 + 2\times1$

    $106 \times 32 = (100+6) \times (30+2) = 100\times30 + 6\times30 + 100\times2 + 6\times2$

3. Using the distributive property, find the following.

    $12 \times 11 = (10+2)\times(10+1) = 10\times10 + 2\times10 + 10\times1 + 2\times1 = 132$

    $13 \times 71 = \boxed{923}$ $104 \times 53 = \boxed{5{,}512}$

    $31 \times 31 = \boxed{961}$ $101 \times 101 = \boxed{10{,}201}$

4. Use the distributive property with subtraction to find the following.

    $19 \times 41 = \boxed{779}$ $19 \times 52 = \boxed{988}$

    $99 \times 71 = \boxed{7{,}029}$ $39 \times 41 = \boxed{1{,}599}$

    $19 \times 19 = \boxed{361}$ $98 \times 99 = \boxed{9{,}702}$

5. Making Connections: Orhan found the smallest pair of twin primes that are greater than 50. He then multiplied them together. What answer did he get if he made a mistake and got 10 more than the correct answer?

    $\boxed{3{,}609}$

6. Challenge: In Tom Brady's first 12 seasons he played 161 games and had an average of 248 yards per game. How many total yards did he pass for? About how many miles is this?

    $\boxed{39{,}928 \text{ yards. This is about } 22.7 \text{ miles}}$

Notes on 5: The connection is to section 7.6. The smallest pair of twin primes greater than 50 are 59 and 61. The product is $(60 - 1) \times (60 + 1) = 3599$. Orhan got 10 more than this. On 6: The total yards is $161 \times 248 = (160 + 1) \times (250 - 2) = (160 \times 250) + (1 \times 250) - (160 \times 2) - (1 \times 2) = 40{,}000 + 250 - 320 - 2 = 40{,}000 - 72 = 39{,}928$. A mile is 1,760 yards. 20 miles is 35,200 yards. Adding 2 more gives 38,720 yards. So the answer is 22 miles and 1,208 yards or about 22.7 miles.

         © 2013 Glenn Ellison

9.3 Patterns in the Multiplication Table

1. Word Problem: John's room is 7 feet by 9 feet. Robert's room is 8 feet by 8 feet. How much larger is Robert's room in square feet?

   > 1 square foot

2. Look at the × 9 column of the multiplication table. Add up the numbers that are shaded the same way in the table below. What answer do you get? Can you explain why using the distributive property?

   > Each pair adds to 99.

   This happens because we shaded pairs of rows that add up to 11. For example, 2 and 9 are shaded the same and $9 \times 2 + 9 \times 9 = 9 \times (2+9) = 9 \times 11$.

3. Look at a 2-by-2 box in the multiplication table that includes two numbers on the main diagonal and two numbers just above and below them. What do you notice about these numbers? How does the sum of the upper left and bottom right numbers compare to the sum of the lower left and upper right numbers? How do the products of the two sets of numbers compare?

   Answers to the first question can vary. The sum of the upper left and lower right is one more than the sum of the other two. The product of the two pairs is the same.

×	1	2	3	4	5	6	7	8	9	10
1	1	2	3	4	5	6	7	8	9	10
2	2	4	6	8	10	12	14	16	18	20
3	3	6	9	12	15	18	21	24	27	30
4	4	8	12	16	20	24	28	32	36	40
5	5	10	15	20	25	30	35	40	45	50
6	6	12	18	24	30	36	42	48	54	60
7	7	14	21	28	35	42	49	56	63	70
8	8	16	24	32	40	48	56	64	72	80
9	9	18	27	36	45	54	63	72	81	90
10	10	20	30	40	50	60	70	80	90	100

9.4 Differences of Squares

1. Word Problem: The ancient Chinese game of Go is played on a 19-by-19 grid. Chess is played on an 8-by-8 board. What is the difference between the number of grid points on a Go board and the number of squares on a chess board?

   $\boxed{297}$

2. Write how you could use the difference of squares formula in these problems.

   $24 \times 26 = (25 - 1) \times (25 + 1) = 25^2 - 1^2$

   $47 \times 53 = (50 - 3) \times (50 + 3) = 50^2 - 3^2$

   $23^2 - 17^2 = (23 + 17) \times (23 - 17) = 40 \times 6$

3. Use the difference-of-squares formula to find the following.

   $7 \times 9 = \boxed{63}$ $6 \times 8 = \boxed{48}$

   $19 \times 21 = \boxed{399}$ $49 \times 51 = \boxed{2499}$

   $9 \times 11 = \boxed{99}$ $8 \times 12 = \boxed{96}$

   $7 \times 13 = \boxed{91}$ $6 \times 14 = \boxed{84}$

4. Use difference-of-squares formula to find the following products and differences.

   $45 \times 55 = \boxed{2475}$ $92 \times 108 = \boxed{9936}$

   $10^2 - 9^2 = \boxed{19}$ $27^2 - 3^2 = \boxed{720}$

   $45^2 - 5^2 = \boxed{2000}$ $99^2 - 1 = \boxed{9800}$

   $76^2 - 14^2 = \boxed{5580}$ $102^2 - 2^2 = \boxed{10400}$

5. Making Connections: Is 851 prime? Hint: Use the difference-of-squares formula.

   $\boxed{\text{NO}}$ (One way to see this is $851 = 30^2 - 7^2 = 37 \times 23$)

6. Challenge: Find the prime factorizations of 273 and 10152.

   $\boxed{273 = 3 \times 7 \times 13}$ $\boxed{10152 = 2^3 \times 3^3 \times 47}$

   Notes on 5: A good way to see this is to note that $851 = 900 - 49 = 30^2 - 7^2 = 37 \times 23$. On 6: $273 = 289 - 16 = 17^2 - 4^2 = 21 \times 13 = 3 \times 7 \times 13$ and $10152 = 10201 - 49 = 101^2 - 7^2 = 108 \times 94 = (4 \times 27) \times (2 \times 47) = 2^3 \times 3^3 \times 47$. Recognizing 10201 as a square is a hard part of the problem, but it did appear in section 9.2 (and will appear again in 14.4).

10.1 How is Area Measured?

1. Word problem: Tim makes a shape using 7 one-foot-by-one-foot floor tiles. If none of the tiles overlap, what is the area of the shape in square feet? Would the answer change if he cut some of the tiles in half to make them triangles instead of squares?

   7 square feet. No, it wouldn't change.

2. In the pictures below suppose that each box was a one foot by one foot square. What would be the area of each shape in square feet?

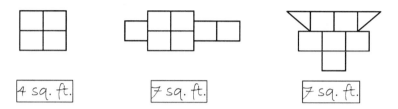

   4 sq. ft.   *7 sq. ft.*   *7 sq. ft.*

3. In the pictures below suppose that each box was a one-centimeter by one centimeter square. What would be the area of each shape in square centimeters?

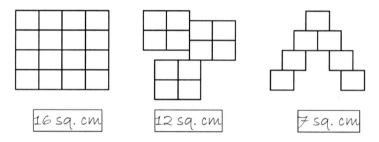

   16 sq. cm   *12 sq. cm*   *7 sq. cm*

4. Making Connections: Suppose that you made a stained glass window by cutting off the corner of one-foot-by-one-foot square pieces of glass to make them into octagons. You then filled in all the gaps with triangles and squares of a different colors and arranged them as shown below. What would be the area of your whole window in square feet?

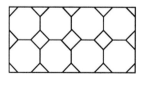

   8 square feet

   Notes on 4: The connection is to section 4.7. To find the area you need to recognize that you don't need to compute the areas of the octagons, squares, and triangles. Instead, note that the area of the total figure doesn't change when you replace the corners with pieces of glass of a different color. The figure started with eight one-foot-by-one-foot squares so the area is 8 square feet.

10.2 Areas of Rectangles

1. Word problem: The front table at a bookstore is 12 feet by 3 feet. What is its area in square feet?

    If a stack of books has a 6 inch by 1 foot footprint, how many stacks can you fit in one square foot? How many stacks of books can be put on the whole table?

    36 sq. ft.   2 stacks fit in one sq. ft.   72

2. Find the area of the rectangles with the following dimensions:

    8" by 8" $8 \times 8 = 64$ sq in   5' by 5'   25 sq. ft.

    7" by 8"   56 sq. in.   10 cm by 12 cm   120 sq. cm

    13 cm by 7 cm 91 sq. cm   24 cm by 3 cm   72 sq. cm

    102" by 12"   1224 sq. in.   34 cm by 51 cm   1734 sq. cm

3. What is the area **in square inches** of a table that is five feet long and two feet wide?

    1440 square inches

4. Making Connections: Use the difference-of-squares formula to find the area in square inches of a rectangle that is 48 inches by 52 inches.

    2496 square inches

5. Challenge: Which has a larger area: a 251 foot by 498 foot backyard or a 249 by 501 foot backyard? What is the positive difference between the areas of the two backyards?

    The 251 by 498 yard is larger by 249 square feet.

Notes on 4: The connection is to section 9.4. The area in square inches is $48 \times 52 = (50 - 2) \times (50 + 2) = 50^2 - 2^2 = 2500 - 4 = 2496$. On 5: It helps to remember the two sided distributive property from section 9.2. The area in square feet of a 249 by 501 backyard is $(250 - 1) \times (500 + 1) = (250 \times 500) - 500 + 250 - 1 = (250 \times 500) - 251$. The area of the 251 by 498 backyard is $(250 + 1) \times (500 - 2) = (250 \times 500) - 500 + 500 - 2 = (250 \times 500) - 2$. The second is larger by 249 square feet.

© 2013 Glenn Ellison

10.3 Areas of Right Triangles

1. Word problem: Griff is remodeling his castle and wants to change the floors in a particular room. The room is in the shape of a right triangle with legs 20 feet and 40 feet. If the cost of a square foot of flooring is 10 copper pennies, how much will Griff have to pay for new flooring material?

   4000 copper pennies

2. Find the area of the right triangles with the following leg lengths:

   8" and 16" $\frac{1}{2} \times 8 \times 16 = 64$ sq in   10 cm and 6 cm   30 sq. cm

   8' and 11'   44 sq. ft.   4' and 21'   42 sq. ft.

   20" and 13"   130 sq. in.   22 cm and 17 cm   187 sq. cm

   101' and 36'   1818 sq. ft.   412 cm and 16 cm   3296 sq. cm

3. Making Connections: Suppose that a right triangle has legs that are 14 feet and 17 feet long. Is its area (in square feet) a prime number?

   No

4. Challenge: If a right triangle has area 28 and integer leg lengths, how many non-congruent triangles are possible?

   4

   Notes on 3: The connection is to section 7.1. The area is ½ × 14 × 17 = 7 × 17 square feet. This shows that it is composite. On 4: Triangles are "congruent" if one can be rotated or reflected to make it match the other. Right triangles will be congruent if and only if the lengths of the legs are the same (even if they are written in a different order). If the area is 28 then the product of the lengths of the two legs is 56. The are four different ways to multiply two positive integers and get 56: 1 × 56, 2 × 28, 4 × 14, and 8 × 7. This gives four non-congruent triangles.

10.4 Areas of Other Triangles

1. Word Problem: Griff has another room that is in the shape of a triangle with base 26 feet and height 10 feet. He wants to floor this one in marble, so the flooring will cost 20 copper pennies per square foot. How much does he have to pay to buy the marble to cover the floor of this room?

   2600 copper pennies

2. Find the area of triangles with the following bases and heights:

   8" and 16"   ½×8×16=64 sq in   4 cm and 7 cm   14 sq. cm

   12' and 7'   42 sq. ft.   21 m and 102 m   1071 sq. m

3. Something that can make finding the areas of triangles tricky is that the "base" is not always on the bottom. In each of the three figures below draw an arrow pointing to the "base" of the triangle. Draw the dashed line that divides the triangle into two right triangles showing the "height". And say what the base and height are.

   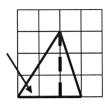   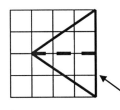   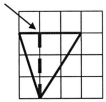

   base = 3   base = 4   base = 3
   height = 3   height = 3   height = 3

4. Making Connections: What is the area of a triangle with base of 99 cm and height of 14 cm?

   693 sq. cm

5. Challenge: If a right triangle has legs 15" and 36" and a hypotenuse with length 39", how long is the perpendicular from the right angle to the hypotenuse?

   13 and 33/39 inches

   Notes on 4: The connection is to section 9.1. The area is ½ × 99 × 14 = 99 × 7 = (100 − 1) × 7 = (100 × 7) − (1 × 7) = 700 − 7 = 693 square feet. On 5: Think about rotating the right triangle so the 39" hypotenuse is the base. The length of "the perpendicular from the right angle to the hypotenuse" is then the height of the triangle. We can compute the area of the triangle in two different ways. One is ½ × 39 × height. The other is as ½ × 15 × 36. So the height is (15 × 36) / 39 = 540 / 39 = 13 and 33/39.

10.5: Finding Areas of Other Shapes: Put Your Addition Skills to Work

1. Draw in dashed lines to show how each of the shapes below could be divided into rectangles and right triangles.

Answers can vary.

2. Draw in dashed lines to divide the shapes below into rectangles and right triangles, and then add up the areas of the parts to find the area of the whole shape.

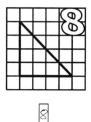 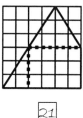 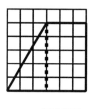 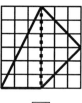

8   21   22½   18

3. Making Connections: An architect designed a tile floor for one of her clients using large octagons and squares that she will have custom cut out of marble. If the picture of the floor below is on graph paper where each square represents one foot, and the three-eighths inch thick marble she is using weighs 5 pounds per square foot, how heavy will each octagonal tile be? How heavy will the diagonally placed square tiles be?

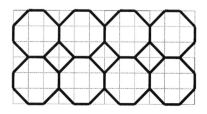

Octagons: 35 lbs
Squares: 10 lbs

4. Challenge: If each square in the grid to the right is one square foot, find the area of the region between the outer hexagon and the inner hexagon.

15 square feet

Notes on 3: Each octagon is 7 square feet and each square is 2 square feet. On 4: The problem can be done either by subtracting the area of the inner hexagon from the area of the outer hexagon or by dividing the area between the two hexagons into a number of triangles and squares. The dashed lines I've drawn in are a good way to divide the larger hexagon into a four triangles and a square. They also show a way to divide the region between the two hexagons into 10 triangles and 2 squares.

Name _____ **Answer Key** _____

10.6: Finding Areas of Other Shapes: Put Your Subtraction Skills to Work

1. Word Problem: Tianxiao wants to make a very boring quilt. It will be a square that is 90 inches on each side. The entire front of the quilt will be blue except for a 30 inch by 30 inch red square in the lower left. How many square inches of blue fabric does Tianxiao need?

$\boxed{7200 \text{ square inches}}$

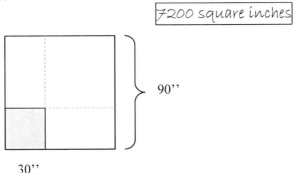

90''

30''

2. You can also find the areas of the shapes below by drawing a bigger rectangle around the shape and subtracting the areas of some right triangles. Draw dashed lines to show how to make a bigger rectangle and draw arrows to show where the triangles are. We've done the first one for you.

3. Find the areas of each of the shapes from question 2 using the subtraction method.

$\boxed{16}$ $\boxed{16^{1}/_{2}}$ $\boxed{16}$ $\boxed{20}$

4. Use addition or subtraction methods to find the areas of each of the shapes below.

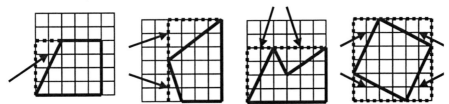

$\boxed{10}$ $\boxed{17}$ $\boxed{14}$ $\boxed{7^{1}/_{2}}$

Notes on 1: The area is $90^2 - 30^2 = (90 - 30) \times (90 + 30) = 60 \times 120 = 7200$ square inches. On 3: The areas are $20 - 4$, $24 - (1.5 + 6)$, $24 - (4 + 4)$, and $36 - (4 \times 4)$. On 4: One way to find the areas pretty easily is to draw the smallest rectangle along the gridlines that encloses each shape and then to subtract off three triangles (and also one rectangle in the case of the rightmost figure).

69

11.1: Counting and Addition

1. Word Problem: Mark went to ski practice at 11 o'clock. Practice lasted two hours. What time did it end?

1 o'clock

2. Fill in the missing numbers to count in mod 12 arithmetic:

   _3_ , 4, 5, 6, _7_ , 8, 9, 10, 11, 12, _1_ , 2, 3, ...

   4, 5, 6, 7, _8_, 9, 10, _11_ , _12_ , _1_ , ...

   _11_ , _12_ , 1, _2_ , _3_ , _4_ , 5, _6_ , 7, ...

3. Find the answers to the following mod 12 addition problems:

   $5 + 5 \equiv 10 \pmod{12}$ $2 + 9 \equiv 11 \pmod{12}$

   $6 + 9 \equiv 3 \pmod{12}$ $7 + 24 \equiv 7 \pmod{12}$

   $6 + 50 \equiv 8 \pmod{12}$ $6 + 6 + 6 \equiv 6 \pmod{12}$

4. What time is it 17 hours after 3 o'clock?

8 o'clock

5. Making Connections: The world's longest debate tournament started at 12 o'clock. The first speaker spoke for one hour. The second spoke for two hours. The pattern continued until the tenth and final speaker spoke for ten hours. What time was it when the last speaker finished speaking?

7 o'clock

6. Challenge: What time is it 390 hours after 2:30?

8:30

Notes on 3: To do $6 + 50$ it's easiest to first note that $50 \equiv 2 \pmod{12}$ so $6 + 50 \equiv 6 + 2 \equiv 8 \pmod{12}$. On 4: Similarly $17 \equiv 5 \pmod{12}$ so 17 hours after 3 o'clock is 5 hours after 3 o'clock on a 12 hour clock. On 5: The connection is to section 3.3. The debate tournament lasted $1 + 2 + 3 + \ldots + 10 = 55$ hours. (Remember that the pairs like 1+10, 2+9, ... each add to 11.) $55 \equiv 7 \pmod{12}$ so the last speaker finishes at 7 o'clock. On 6: A good way to simplify large numbers is to subtract off whatever multiple of 12 you recognize. Here, a natural first step is to note that 360 is a multiple of 12 so $390 \equiv 30 \equiv 6 \pmod{12}$. Six hours after 2:30 is 8:30.

Name _____**Answer Key**_____

11.2: Hard Math for Clock-World Elementary School

1. Word Problem: Kate got up at 7 o'clock. One hour later she left for school. She stayed there for 3 hours. Then she got picked up early by her dad. They drove for nine hours. And then Kate slept for 11 hours. What time did Kate wake up on the second day?

     7 o'clock

2. Mod 12 multiplication can be easier than multiplication in our world because there are more patterns. Find patterns and use them to answer the following problems:

    $1 \times 6 \equiv 6$ (mod 12) $2 \times 6 \equiv 12$ (mod 12) $3 \times 6 \equiv 6$ (mod 12)

    $4 \times 6 \equiv 12$ (mod 12) $5 \times 6 \equiv 6$ (mod 12) $6 \times 6 \equiv 12$ (mod 12)

    $7 \times 6 \equiv 6$ (mod 12) $10 \times 6 \equiv 12$ (mod 12) $100 \times 6 \equiv 12$ (mod 12)

3. Solve the following mod 12 multiplication problems.

    $1 \times 4 \equiv 4$ (mod 12) $2 \times 4 \equiv 8$ (mod 12) $3 \times 4 \equiv 12$ (mod 12)

    $4 \times 4 \equiv 4$ (mod 12) $5 \times 5 \equiv 1$ (mod 12) $6 \times 6 \equiv 12$ (mod 12)

    $7 \times 7 \equiv 1$ (mod 12) $3 \times 9 \equiv 3$ (mod 12) $13 \times 13 \equiv 1$ (mod 12)

4. What is $11 \times 11 \times 11 \times 11 \times 11$ in mod 12 arithmetic?

     11

5. Making Connections: Use the distributive property to simplify $(14 \times 5) + (14 \times 8)$ in mod 12 arithmetic?

     2

6. Making Connections: What is $1 + 2 + 3 + 4 + 5 + \ldots + 11$ in mod 12 arithmetic?

     6

7. Challenge: Lucy's clock runs fast and gains 1 hour each day. If she sets the clock to the correct time at noon on January 1, 2012, what time will the clock say at noon on January 1, 2013?

    6 pm

    Notes on 5: The connection is to section 9.1. $(14 \times 5) + (14 \times 8) = 14 \times (5 + 8) = 14 \times 13 \equiv 14 \times 1 \equiv 2$ (mod 12). On 6: The connection is to section 3.3. Each pair like $1 + 11$ adds to $12 \equiv 0$ (mod 12) so the mod 12 sum is just the number in the middle, 6. On 7: The clock gains 366 hours during 2012 (because it is a leap year and has 366 days). $366 \equiv 6$ (mod 12).

11.3: Mod 10 Arithmetic

1. Word Problem: A common way of picking one kid out of a group is to point to the kids in order and say "Eeny , meeny, miny, moe, Catch a tiger …". When you do this you point to the kid #1 when saying "Ee", kid #2 when saying "ny", and continue to go around the circle pointing to a new kid on each syllable. The whole rhyme has 28 syllables. If you use this method to choose one kid from a group of 12, the fourth kid will be chosen because $28 \equiv 4 \pmod{12}$. Which kid would be chosen from a group of 10? How about from a group of 3?

    The 8th kid will be chosen from a group of 10 and the 1st from a group of 3.

2. Fill in the missing numbers to count in mod 10 arithmetic:

    1, 2, _3_ , 4, 5, 6, _7_ , 8, 9, _0_ , 1, 2, 3, …

    4, 5, 6, 7, _8_ , 9, 0, _1_ , _2_ , …

    _3_ , _4_ , 5, 6, , _7_ , _8_ , _9_ , 0, _1_ , 2, …

3. Find the answers to the following mod 10 addition problems:

    $5 + 7 \equiv 2$ (mod 10) $2 + 9 \equiv 1$ (mod 10)

    $8 + 9 \equiv 7$ (mod 10) $9 + 6 \equiv 5$ (mod 10)

4. Find the answers to the following mod 10 multiplication problems:

    $1 \times 7 \equiv 7$ (mod 10) $0 \times 9 \equiv 0$ (mod 10)

    $8 \times 9 \equiv 2$ (mod 10) $9 \times 6 \equiv 4$ (mod 10)

    $27 \times 5 \equiv 5$ (mod 10)

5. Making Connections: Use the two-sided distributive property to write 106×37 as a sum of four terms. What are the values of each of the four terms in mod 10 arithmetic?

    $106 \times 37 = 100 \times 30 + 6 \times 30 + 100 \times 7 + 6 \times 7$. The values are 0, 0, 0, and 2.

6. Challenge: What is the pattern when you keep multiplying 9 by itself in mod 10 arithmetic? What is the value of 9^9 in mod 10 arithmetic? What is the last digit of 9^{9^9} when it is written out as a base 10 number?

    They alternate 9, 1, 9, 1, …   9   9

    Notes on 1: The 8th is chosen from a group of 10 because $28 \equiv 8 \pmod{10}$. On 5: The connection is to section 9.2. On 6: If you start with 9 and keep multiplying by 9 in mod 10 arithmetic the numbers go 9, 1, 9, 1, … One can also think of this sequence as $9^1, 9^2, 9^3, 9^4$, … . 9^9 would be the ninth term in this sequence, so it is equal to 9. 9^{9^9} would be the 9^9th term in this sequence. In general all odd terms are equal to 9 and all even terms are equal to 1. 9^9 is an odd number so the 9^9th term would be 9.

11.4: Divisibility Rules

1. Underline all of the multiples of 3 in the array below. (The first row is already done for you.)

1	12	23	34	45
27	15	29	10	99
33	333	222	4444	55555

2. Underline all of the multiples of 9 in the array below. (The first row is already done for you.)

1	12	23	34	45
27	215	99	106	699
22	333	4444	55555	777777777

3. Siyuan added together two numbers and got an answer that was a multiple of 9. The two numbers he added were each between 50 and 55. What was the answer that he got?

    108

4. Making Connections: Use your knowledge of prime factorization to figure out which of 1, 10, 100, 1000, 10000, … are multiples of 4?

   Every number from 100 on is a multiple of 4.

5. What is the value of 1000 + 300 + 50 + 8 in mod 4 arithmetic? Is 1358 a multiple of 4?

    2   No

6. Challenge: Is 98,765,432 a multiple of 8?

    Yes

Notes on 2: A number is a multiple of 9 if the sum of the digits is a multiple of 9. On 3: The sum will be between 100 and 110. The only multiple of 9 in this range is 108. On 4: The (loose) connection is to section 7.4. $10 = 2 \times 5$, so $10^2 = 2^2 \times 5^2$, and this and all later terms are multiples of 4. On 5: One way to simplify 1358 in mod 4 arithmetic is to simplify each term separately. $1000 \equiv 0$ (mod 4). $300 \equiv 0$ (mod 4). $50 \equiv 2$ (mod 4). $8 \equiv 0$ (mod 4). So the sum is equal to 2 (mod 4). On 6: By reasoning similar to that in problem 4 we know that 1000, 10000, and all higher powers of 10 are multiples of 8. So evaluating the number in a similar manner to problem 5 we find that $98,765,432 = 90,000,000 + 8,000,000 + \dots + 5,000 + 400 + 30 + 2 \equiv 0 + 0 + \dots + 0 + 400 + 30 + 2 \equiv 0 + 6 + 2 \equiv 0$ (mod 8).

12.1: Multiplying By One Digit: The Partial Products Method

1. Word Problem: If you eat three meals every day, how many meals will you eat in the year 2017?

   $\boxed{1,095}$

2. Fill in the missing numbers in the calculations below to complete the partial-products multiplications.

$$
\begin{array}{r}
135 \\
\times\ \ 7 \\
\hline
700 \\
210 \\
35 \\
\hline
945
\end{array}
\qquad
\begin{array}{r}
2830 \\
\times\ \ \ 4 \\
\hline
8000 \\
3200 \\
120 \\
\hline
11320
\end{array}
\qquad
\begin{array}{r}
398 \\
\times\ \ 8 \\
\hline
2400 \\
720 \\
64 \\
\hline
3184
\end{array}
$$

3. Use the partial products method to do each of the multiplications below.

$$
\begin{array}{r}
244 \\
\times\ \ 4 \\
\hline
\boxed{976}
\end{array}
\qquad
\begin{array}{r}
830 \\
\times\ \ 7 \\
\hline
\boxed{5,810}
\end{array}
\qquad
\begin{array}{r}
695 \\
\times\ \ 3 \\
\hline
\boxed{2,085}
\end{array}
\qquad
\begin{array}{r}
287 \\
\times\ \ 5 \\
\hline
\boxed{1,435}
\end{array}
$$

$$
\begin{array}{r}
1267 \\
\times\ \ \ 4 \\
\hline
\boxed{5,068}
\end{array}
\qquad
\begin{array}{r}
3087 \\
\times\ \ \ 7 \\
\hline
\boxed{21,609}
\end{array}
\qquad
\begin{array}{r}
3737 \\
\times\ \ \ 9 \\
\hline
\boxed{33,633}
\end{array}
\qquad
\begin{array}{r}
39594 \\
\times\ \ \ \ 6 \\
\hline
\boxed{237,564}
\end{array}
$$

4. Challenge: The number 123 can be written as $111 + 11 + 1$. Use a similar expansion (and the distributive property) to find 123456789×9 and then check your answer using the partial products method.

   $\boxed{1,111,111,101}$

Notes on 1: The answer is 365×3. On 4: This problem makes a couple connections. Write 123456789×9 as $(111,111,111 + 11,111,111 + \ldots + 111 + 11 + 1) \times 9$. Using the distributive property as in section 9.1 this is $999,999,999 + 99,999,999 + \ldots + 999 + 99 + 9$. One could add this using addition with carrying. But an even easier way is to use the making round numbers technique from section 3.4. The sum is $(1,000,000,000 - 1) + (100,000,000 - 1) + \ldots + (1,000 - 1) + (100 - 1) + (10 - 1)$. There are nine -1's in this expression so the answer is $1,111,111,110 - 9 = 1,111,111,101$.

12.2: One-Digit-One-Line Long Multiplication

1. Word Problem: Nicholas has 123 nickels. Nicole has 200 nickels. And Nikolai has 34. If they go to the bank together and ask to get pennies for all of their nickels, how many pennies will they get?

   1,785 pennies

2. Fill in the missing steps in the calculations below to show how you would compute 259×6 and 837×5 using one-digit-one-line multiplication

```
              6            4 6        1 4 6      1 4 6
    259        259         259         259        259
  ×   7   →  ×   7   →   ×   7   →   ×   7   →  ×   7
  _____      _____       _____       _____      _____
                 3          13         813       1813
```

```
              3            1 3        4 1 3      4 1 3
    837        837         837         837        837
  ×   5   →  ×   5   →   ×   5   →   ×   5   →  ×   5
  _____      _____       _____       _____      _____
                 5          85         185       4185
```

3. Do each of the multiplications below using one-digit-one-line multiplication. This time just do the multiplications and don't write anything other than the answer and the small numbers you use to keep track of the carrying.

```
    3 4           1 1           1 1           4 3
     47            22            65            87
  ×   7         ×   7         ×   3         ×   5
  _____         _____         _____         _____
   329           154           195           435
```

```
      2             5           1 1           1 1
     17            80            66            33
  ×   4         ×   7         ×   3         ×   6
  _____         _____         _____         _____
    68           560           198           198
```

```
    1 2             2          1 2 2         7 6 4
    305           510           123           886
  ×   4         ×   4         ×   9         ×   8
  _____         _____         _____         _____
   1220          2040          1107          7088
```

Notes on 1: The total number of nickels is $123 + 200 + 34 = 357$. Multiplying 357×5 gives 1,785.

12.2: One-Digit-One-Line Long Multiplication

4. Do each of the multiplications below using one-digit-one-line multiplication.

^{2 3 4}
  347
×   7

2429

  830
×   7

5810

  126
×   8

1008

  287
×   5

1435

  1267
×   4

5068

  3087
×   7

21609

  3737
×   9

33633

  39594
×   6

237564

  3055
×   4

12220

  5117
×   7

35819

  9876
×   4

39504

  42059
×   8

336472

  23056
×   3

69168

  56136
×   6

336816

  90023
×   8

720184

  50305
×   7

352135

5. Making Connections: What is the sum of the measures of all of the interior angles of a regular octagon?

1080°

6. Challenge: What is 142,857,285,714,142,857 × 7?

1,000,000,999,998,999,999

Notes on 5: The connection is to section 4.4. Each angle in a regular octagon is 135 degrees. So the sum of the measures of all the angles is $135 \times 8 = 1080°$. Another formula that some students may know is that the sum of the interior angles of any n-gon is $(n-2) \times 180°$. So the sum could alternately be computed as 180×6. On 6: One way to get the answer is with a straight application of one-digit-one-line long multiplication. An easier way is to recognize a pattern in big number: $142,857,285,714,142,857 = 142,857 \times (1,000,000,000,000 + 2,000,000 + 1)$. So after computing $142,857 \times 7 = 999,999$ you can find the answer using the distributive property as $999,999,000,000,000,000 + 1,999,998,000,000 + 999,999$.

12.3: Long Multiplication

1. Word Problem: Sandor hiccupped 23 times per minute for an entire hour. How many hiccups is this?

    $\boxed{1380}$

2. In each of the long multiplication problems below I multiplied the number on the top by the ones digit for you. Finish the multiplications by multiplying the number on the top by the tens digit, putting the product in the right place, and adding.

    ```
        2 3              2                2 1 5
       35             106                317
     × 17           ×  24              ×  18
      245            424               2536
      35             212               317
      595           2544               5706
    ```

3. In each of the long multiplication problems below I multiplied by both the tens digit and the hundreds digit for you. But I didn't multiply by the ones digit. Finish the problems by multiplying the number on top by the ones digit, putting the product in the right place, and adding.

    ```
       46             170               355
     × 371          × 672             × 542
       46             340               710
      322           1190              1420
      138           1020              1775
    17066         114240            192410
    ```

4. Challenge: Think about how long multiplication would work in base 8 and try to use it to compute answers to the base 8 multiplication problems below. Give the answers in base 8.

    ```
       23 (8)          34 (8)            33 (8)
     × 11 (8)        × 12 (8)          × 32 (8)
       23              70                66
       23              34               121
      253 (8)         430 (8)          1276 (8)
    ```

Notes on 1: The answer is 23×60. On 4: Students need to remember section 2.6 well, and even then this will be confusing. One thing to remember is that each one-digit multiplication is a base 8 multiplication. They can be done with repeated addition using base 8 carrying, e.g. $34_{(8)} \times 2_{(8)} = 34_{(8)} + 34_{(8)} = 70_{(8)}$. Or, more quickly but confusingly, one can do them using the base 8 version of partial products or one-digit-one-line multiplication, e.g. $33_{(8)} \times 3_{(8)} = 110_{(8)} + 11_{(8)} = 121_{(8)}$. Or one can convert to base 10 and convert back, e.g. $33_{(8)} \times 3_{(8)} = 27_{(10)} \times 3_{(10)} = 81_{(10)} = 121_{(8)}$. And then at the end, you need to add the columns using base 8 addition with carrying.

12.3: Long Multiplication

5. Use long multiplication to find the answers to each of the following problems.

```
  3 4
   47              23              65              34
 × 17           × 11            × 11            × 21
  329            253             715             714
  47
  799
```

```
   17              80              45             133
 × 44           × 37            × 31            × 14
  748           2960            1395            1862
```

```
  305             510             123             886
 × 24           × 84           × 101           × 113
 7320           42840           12423          100118
```

```
  225             307            1024             816
 × 124          × 714          × 1024           × 373
27900          219198         1048576          304368
```

6. Making Connections: Supposed that on a multiple choice test you were asked to find 3874 × 37 and given the choices below. How could you quickly pick the right answer?

   A) 16,248   B) 113,456   C) 141,639   D) 143,338   E) 1,464,646

   The answer is D). The last digit is 8. And it must be over 100,000.

Notes on 6: The connection is to section 11.3. Using mod 10 arithmetic (or just thinking about the last digit) we know that the number ends with an 8. To figure out about how big the answer will be one can approximate the multiplication is various ways, e.g. it is about 4000 × 35 = 140,000, or just note that choice A) 16,248 is much too small because 3874 × 10 = 38,740 is already much bigger than this.

13.1: Making Ordered Lists

1. Word Problem: The accounts of ancient Egyptian poets indicate that Cleopatra died after being bitten by an asp (also known as an Egyptian cobra). How many three letter English words can you make by rearranging the letters in the word ASP? (Include ASP.)

   3

2. List all two digit numbers that use only the digits 1, 3, and 5. Put the numbers starting with 1 first, then those starting with 3, then those starting with 5.

   11, 13, 15, 31, 33, 35, 51, 53, 55

3. List all three digit numbers that use only the digits 0 and 1.

   100, 101, 110, 111

4. List all three-letter sequences you can make using the letters A and D in the order that they would appear in the dictionary. How many are English words?

   AAA, AAD, ADA, ADD, DAA, DAD, DDA, DDD. Two are words.

5. At the start of a Scrabble game Adrienne had four A's, a T, an N, and an O. She started the game by making a two-letter English word. List all of the words that she might possibly have made.

   AA, AN, AT, NA, NO, ON, TA, TO

6. Suppose that Jack is taller than Kevin, who is taller than Penelope, who is taller than Kate. Ms Cronin wants to pick one of four to be the narrator for a class play and one other to be the director. She wants the director to be taller than the narrator. In how many ways can she assign these two jobs?

   6

7. Making Connections: How many three-digit prime numbers can be made using only the digits 0 and 1.

   1 (Only 101 is prime.)

8. Challenge: Suppose that there are five kids on the Cabot math team: Anna, Bob, Clarence, David and Evelyn. The coach wants to choose three kids to go to a meet. Bob will only go if David is also going. Anna won't go if Clarence is going. David won't go if he's the only boy. In how many ways can the team be chosen?

   4 ways. (ABD, BCD, BDE, CDE)

Notes on 1: There are six arrangements. ASP, SAP, and SPA are words. On 4: ADD and DAD are words. On 5: The question is a bit unfair in that the official Scrabble dictionary includes many rarely used "words" including AA, NA, and TA in addition to the five more normal words. (There are 13 two-letter sequences.) On 7: The connection is to sections 7.3 and 7.5. There are 4 possible numbers 100, 101, 110, and 111. Only 101 is prime. On 8: One way is just to write out the 10 possible combinations and then cross out the ones that violate each rule.

13.2: Counting Without Listing: Multiply the Number of Choices

1. Word Problem: The apartment "numbers" in a building are actually two character codes: the first is a number from 1-3 indicating which floor the apartment is on and the second is a letter from A to G indicating which apartment on the floor it is. For example, 2C is one of the apartments on the 2nd floor. How many apartments are in the building?

    21

2. How many ways are there to make a two-character code consisting of a letter chosen from A, B, C, D, E followed by a digit chosen from 1 to 5? (For example, A3 and B4 are two possible codes.)

    25

3. How many three-character codes like A7X and B3Z can you make if the first character must be A, B, C, D, or E, the second must be a digit chosen from 0-9, and the third can be either X or Z.

    100

4. There are 23 students in Ms. Ninorc's class and 22 students in Mr. Ametlub's. The teachers want to choose one student from each class to have ice cream with Principal Skinner. How many ways are there to choose the pair of students?

    506

5. How many three-digit odd numbers are there?

    450

6. Making Connections: In Massachusetts the current Red Sox license plates all start with RS. They then have two digits chosen from 0-9 and then they have two letters. Use the distributive property (or long multiplication) to figure out how many such license plates are possible.

    67,600

7. Challenge: You can think of a digital clock as a machine that displays a three or four digit number in each minute. How many different numbers are displayed during the course of the day?

    720

    (or 1440 if you're in a country where clocks read 19:30 at 7:30pm)

Notes on 4: There are 23 × 22 choices. On 5: There are 9 choices for the first digit (it can't be 0), 10 choices for the second digit, and 5 choices for the 3rd digit. 9 ×10 × 5 = 450. The connection is to sections 9.1 or 12.3. The answer is 26 × 26 × 10 × 10. Students can do 26 × 26 using long multiplication or as (25 + 1) × (25 + 1). On 7: The easiest way to think of this is that there are 12 possible choices for the "hour" part (1, 2, …, 12) and then 60 choices for the choices for the "minute" part of the display.

13.3: Counting Without Listing: Orderings and Choices that Change

1. Word Problem: Mr. Zhu is having three students give oral presentations: Anna, Catherine, and Ethan. List all of the different orders in which he could have them give their presentations.

   Anna, Catherine, Ethan   *Catherine, Anna, Ethan*   *Ethan, Anna, Catherine*
   Anna, Ethan, Catherine   *Catherine, Ethan, Anna*   *Ethan, Catherine, Anna*

2. Suppose you are given five number tiles: 1, 2, 3, 5, and 7. How many different 2-digit numbers can you make by choosing two of these tiles?

    20

3. How many four-letter sequences can you make by rearranging the letters in the word MEAT? (Include MEAT in your count.)

    24

4. There are six schools in a math contest. At the end of the contest they want to give the winning team a plaque that has the first-place, second-place, and third-place winners engraved on it. For example, they might want it to say 1st Place – Underwood, 2nd Place – Cabot, 3rd Place – Mason Rice. This is hard to do, however, because it takes a long time to engrave names on the plaque. Mr. Martian has an idea: he thinks that they could just order plaques in advance showing every possible order and then pick the right one to give out when they see the results. How much would it cost to do this if each plaque costs $10?

   $1,200

5. Making Connections: A multiple choice test asked how many three letter codes there are in which all three letters are different. It gave the choices below. Iris used mod 10 arithmetic to quickly find the answer. What is it?

   (A) 75   (B) 676   (C) 15,600   (D) 17,576   (E) 32,768

    (C) 15,600

6. Challenge: How many 5-digit odd numbers can you make by rearranging the digits of the number 12354? (Hint: Think about picking the ones digit first.)

    72

Notes on 2: There are $5 \times 4 = 20$ choices. On 3: There are $4 \times 3 \times 2 \times 1 = 24$ choices. On 4: They would need to order $6 \times 5 \times 4 = 120$ plaques. On 5: The connection is to section 11.3. The number of three letter codes is $26 \times 25 \times 24$. In mod 10 arithmetic this can be computed as $6 \times 5 \times 4 \equiv 0 \times 4 \equiv 0 \pmod{10}$ so she knew that the correct answer would end with a 0. On 6: One good way to do this is to think about choosing the digits from right to left. There are 3 choices for the one's digit (1, 3, or 5). And then you have 4 choices left for the tens digit, 3 for the hundreds digit, 2 for the thousands digit, and 1 for the ten thousands. $3 \times 4 \times 3 \times 2 \times 1 = 72$.

13.4: Choosing Groups of Two: Undo the Double Counting

1. Word Problem: Mr. Zhu has three students who need to give oral presentations: Anna, Catherine, and Ethan. But there is only enough time for two of them to speak. List all of the different groups of two students that he can pick.

    > Anna, Catherine   Anna, Ethan   Catherine, Ethan

2. Alex, Beata, Carter, and Dai are walking home from school. If they want to pick one person to go back to the playground to get something they forgot, how many people can they pick? If they then want to pick a second person to go with the first person how many people can they then choose from? How many ways are there make the two choices? Does each different choice produce a different group of two kids walking back to the playground?

    > 4, 3, 12, No

3. Gino's Pizza offers 8 possible pizza toppings. If they tried to list all possible two-topping pizzas on their menu, how many different pizzas would be on the list?

    > 28

4. Making Connections: Suppose there are 202 7th graders at Bigelow. The school needs to pick two kids to represent it at the state geography bee. In how many different ways can it pick the two students?

    > 20,301

5. Challenge: There are 5 teams in the AL East: Baltimore, Boston, New York, Tampa, and Toronto. In the old MLB playoff system sometimes only one of the five teams would make the playoffs, and in other years a second team would also make the playoffs as a "wild card". How many different possibilities were there for the set of AL East teams that make the playoffs?

    > 15

Notes on 3: There are 8 choices for the first topping, and 7 for the second, but this is double counting so the answer is $(8 \times 7) \div 2 = 28$. On 4: The connection is to section 9.1. The number of ways to choose the two kids is $(202 \times 201) \div 2 = 101 \times 201$. A good way to compute this is with the distributive property: $201 \times (100 + 1) = (201 \times 100) + (201 \times 1) = 20,100 + 201 = 20,301$. On 5: A good way to do this is to count separately the number of ways without a wildcard from the AL East (this is just 5) and the number of ways with a wildcard from the AL East (this is $5 \times 4 \div 2$).

14.1: Squaring Small Numbers: Just Memorize the Answers

1. Word Problem: How many squares are there on an 8 by 8 checkerboard?

    64

2. Write down the answers to each of the multiplication problems below.

    $1 \times 1 = 1$ $2 \times 2 = 4$ $3 \times 3 = 9$ $4 \times 4 = 16$

    $5 \times 5 = 25$ $6 \times 6 = 36$ $7 \times 7 = 49$ $8 \times 8 = 64$

    $9 \times 9 = 81$ $7 \times 7 = 49$ $5 \times 5 = 25$ $4 \times 4 = 16$

3. Write down the answers to each of the squaring problems below.

    $11^2 = 121$ $12^2 = 144$ $13^2 = 169$ $14^2 = 196$

    $15^2 = 225$ $12^2 = 144$ $11^2 = 121$ $12^2 = 144$

    $13^2 = 169$ $14^2 = 196$ $11^2 = 121$ $15^2 = 225$

    $12^2 = 144$ $11^2 = 121$ $15^2 = 225$ $13^2 = 169$

4. Write down the answers to each of the squaring problems below.

    $16^2 = 256$ $17^2 = 289$ $18^2 = 324$ $19^2 = 361$

    $20^2 = 400$ $16^2 = 256$ $18^2 = 324$ $19^2 = 361$

5. Making Connections: How many square inches are there in one square foot?

    144

6. Challenge: Write out the squares of all of the whole numbers in the 20's.

    $20^2 = 400$ $21^2 = 441$ $22^2 = 484$ $23^2 = 529$ $24^2 = 576$
    $25^2 = 625$ $26^2 = 676$ $27^2 = 729$ $28^2 = 784$ $29^2 = 841$

Notes on 5: The connection is to section 10.2. There are $12 \times 12 = 144$ square inches in a square foot. On 6: Students can compute these squares using the one-sided or two-sided distributive property or using long multiplication.

Name _____**Answer Key**_____

14.2: Squaring Bigger Numbers: Slide Apart, Multiply, Add Back the Square

1. Fill in the blanks to make each of the following slide-multiply-and-add formulas true.

    $10 \times 10 = (9 \times 11) + \underline{\ 1\ }$ $13 \times 13 = (12 \times \underline{14}) + 1$

    $10 \times 10 = (8 \times 12) + \underline{\ 4\ }$ $13 \times 13 = (11 \times \underline{15}) + 4$

    $10 \times 10 = (7 \times 13) + \underline{\ 9\ }$ $13 \times 13 = (\underline{10} \times 16) + 9$

2. Write out slide-multiply-and-add formulas that you could use to square the numbers below using a multiplication by 10.

    $14 \times 14 = (\underline{10 \times 18}) + 4^2$ $13 \times 13 = (\underline{10 \times 16}) + 3^2$

    $12 \times 12 = (\underline{10 \times 14}) + 2^2$ $11 \times 11 = (\underline{10 \times 12}) + 1^2$

3. Now do the multiplications and additions.

    $14 \times 14 = \underline{180 + 16 = 196}$ $13 \times 13 = \underline{160 + 9 = 169}$

    $12 \times 12 = \underline{140 + 4 = 144}$ $11 \times 11 = \underline{120 + 1 = 121}$

4. Use the slide-multiply-and-add trick to square the numbers below.

    $21 \times 21 = \underline{(20 \times 22) + 1 = 441}$ $22 \times 22 = \underline{(20 \times 24) + 4 = 484}$

    $23 \times 23 = \underline{(20 \times 26) + 9 = 529}$ $26 \times 26 = \underline{(22 \times 30) + 16 = 676}$

    $19 \times 19 = \underline{(18 \times 20) + 1 = 361}$ $29 \times 29 = \underline{(28 \times 30) + 1 = 841}$

5. The slide-multiply-and-add trick works is particularly easy to use when numbers end in 5 – the answers always end in 25 and the start of the number has an easy-to-learn pattern. Use this pattern to fill in the rest of the answers below.

    $25 \times 25 = \underline{625}$ $35 \times 35 = \underline{1225}$ $45 \times 45 = \underline{2025}$

    $55 \times 55 = \underline{3025}$ $65 \times 65 = \underline{4225}$ $75 \times 75 = \underline{5625}$

    $85 \times 85 = \underline{7225}$ $95 \times 95 = \underline{9025}$ $15 \times 15 = \underline{225}$

    $25 \times 25 = \underline{625}$ $55 \times 55 = \underline{3025}$ $75 \times 75 = \underline{5625}$

    $35 \times 35 = \underline{1225}$ $85 \times 85 = \underline{7225}$ $55 \times 55 = \underline{3025}$

6. Challenge: Find 987^2.

    $\boxed{974,169}$

Notes on 6: Sliding the numbers apart by 13 gives $987 \times 987 = (974 \times 1,000) + 13^2 = 974,000 + 169 = 974,169$.

84

14.3: The Big Slide

1. If you're trying to compute 24^2, you can use the slide, multiply, and add back the square method in several different ways. You could compute $(23 \times 25) + 1^2$. You could compute $(20 \times 28) + 4^2$. You could compute $(18 \times 30) + 6^2$. Or you could compute $(10 \times 38) + 14^2$. All, of course, give the same answer. Write down several ways to compute 22^2 and see which one you think is easiest.

    Answers can vary. Some ways are:

    $(20 \times 24) + 2^2, (19 \times 25) + 3^2, (14 \times 30) + 8^2, (10 \times 34) + 12^2$

2. Fill in the blanks to show how you could use the big slide to square each of the following numbers.

    $97^2 = (94 \times 100) + \underline{3^2}$ $94^2 = (\underline{88} \times 100) + 6^2$

    $91^2 = (82 \times \underline{100}) + 9^2$ $89^2 = (78 \times 100) + \underline{11^2}$

    $83^2 = (66 \times 100) + \underline{17^2}$ $99^2 = (\underline{98} \times 100) + 1^2$

3. Use the big slide to square each of the following numbers.

    $99^2 = (98 \times 100) + 1^2 = 9801$ $98^2 = (96 \times 100) + 2^2 = 9604$

    $97^2 = (94 \times 100) + 3^2 = 9409$ $94^2 = (88 \times 100) + 6^2 = 8836$

    $91^2 = (82 \times 100) + 9^2 = 8281$ $89^2 = (78 \times 100) + 11^2 = 7921$

    $83^2 = (66 \times 100) + 17^2 = 6889$ $85^2 = (70 \times 100) + 15^2 = 7225$

4. Use a big slide to square each of the following numbers.

    $78^2 = (56 \times 100) + 22^2 = 5600 + 484 = 6084$

    $75^2 = (50 \times 100) + 25^2 = 5000 + 625 = 5625$

    $69^2 = (38 \times 100) + 31^2 = 3800 + 961 = 4761$

5. Making Connections: Find 111^2 using both the distributive property and the big slide method.

    $111^2 = (100 \times 111) + (10 \times 111) + (1 \times 111) = 11,100 + 1,110 + 111 = 12,321$

    $111^2 = (100 \times 122) + 11^2 = 12,200 + 121 = 12,321$

6. Challenge: You can use the slide method twice to find 1102^2. Try it.

    $1102^2 = (1000 \times 1204) + 102^2 = 1,204,000 + (100 \times 104) + 2^2 = \boxed{1,214,404}$

    Notes on 5: The distributive-property calculation is simplest if you just use the one-sided distributive property of section 9.1 expanding one of the 111's as $(100 + 10 + 1)$. On 6: The first slide is to slide the numbers apart by 102: $1102 \times 1102 = (1000 \times 1204) + 102^2$. The first product is 1,204,000. And then the second term can be evaluated by sliding apart by 2: $102 \times 102 = (100 \times 104) + 2^2 = 10,404$.

14.4: Another Distributive Identity: $(a + b)^2 = a^2 + 2ab + b^2$

1. Show how to use the formula above to solve the following squaring problems.

    $(10 + 3)^2 = 10^2 + 2 \times 10 \times 3 + 3^2$

    $(20 + 1)^2 = 20^2 + 2 \times 20 \times 1 + 1^2$

    $(30 + 1)^2 = 30^2 + 2 \times 30 \times 1 + 1^2$

    $(50 + 2)^2 = 50^2 + 2 \times 50 \times 2 + 2^2$

2. Now compute the squares using the $(a + b)^2$ formula.

    $13^2 = (10 + 3)^2 = 100 + 60 + 9 = 169$

    $21^2 = (20 + 1)^2 = 400 + 40 + 1 = 441$

    $31^2 = (30 + 1)^2 = 900 + 60 + 1 = 961$

    $104^2 = (100 + 4)^2 = 10,000 + 800 + 16 = 10,816$

3. Show how you could use the formula $(a - b)^2 = a^2 - 2ab + b^2$ to compute the squares below.

    $(10 - 3)^2 = 10^2 - 2 \times 10 \times 3 + 3^2$

    $(20 - 1)^2 = 20^2 - 2 \times 20 \times 1 + 1^2$

    $(50 - 2)^2 = 50^2 - 2 \times 50 \times 2 + 2^2$

4. Now compute each of the squares using the distributive identity.

    $7^2 = (10 - 3)^2 = 100 - 60 + 9 = 49$

    $19^2 = (20 - 1)^2 = 400 - 40 + 1 = 361$

    $48^2 = (50 - 2)^2 = 2500 - 200 + 4 = 2304$

    $96^2 = (100 - 4)^2 = 10,000 - 800 + 16 = 9216$

5. Making Connections: Daesun's back yard is 100 feet by 100 feet. There is a 3 foot wide gravel border around the outside. The rest of the yard is planted with grass. How large is the grassy area in square feet?

    8836 square feet

6. Challenge: Using the distributive property see if you can figure out a good way to compute 17^4.

    One way is $17^4 = 289^2 = (300 - 11)^2 = 90,000 - 6,600 + 121 = 83,521$

    Notes on 5: The connection is to section 10.2. The part of the yard inside the border is just a 94 foot by 94 foot square. (You take off 3 feet on each side). So the area is $94^2 = 8836$ square feet. On 6: A good first step is to recognize that $17^4 = (17^2 \times 17^2) = 289 \times 289$. The distributive property is relatively easy to use in this problem because multiplying by 11 is relatively easy. $(300 - 11) = 300^2 - (2 \times 300 \times 11) + 11^2$.

© 2013 Glenn Ellison

14.5: Squaring Three Digit Numbers

1. Use the big slide method to square each of the following numbers.

    $112^2 = (100 \times 124) + 12^2 = 12,400 + 144 = 12,544$

    $105^2 = (100 \times 110) + 5^2 = 11,000 + 25 = 11,025$

    $121^2 = (100 \times 142) + 21^2 = 14,200 + 441 = 14,641$

    $206^2 = (200 \times 212) + 6^2 = 42,400 + 36 = 42,436$

2. Use $(a + b)^2$ formula to compute each of the following.

    $112^2 = 100^2 + 2 \times 100 \times 12 + 12^2 = 10,000 + 2,400 + 144 = 12,544$

    $105^2 = 100^2 + 2 \times 100 \times 5 + 5^2 = 10,000 + 1,000 + 25 = 11,025$

    $121^2 = 100^2 + 2 \times 100 \times 21 + 21^2 = 10,000 + 4,200 + 441 = 14,641$

    $206^2 = 200^2 + 2 \times 200 \times 6 + 6^2 = 40,000 + 2,400 + 36 = 42,436$

3. Use the big slide method to compute 993^2.

    $\boxed{986,049}$

4. Try to compute 512^2 in your head.

    $\boxed{262,144}$

5. Making Connections: How many possible license plates are there if each license plate has two letters followed by four digits?

    $\boxed{6,760,000}$

6. Challenge: Using the fact that $2^{10} = 1024$, find 2^{20}.

    $\boxed{1,048,576}$

Notes on 3: Sliding the numbers apart by 7 gives $993 \times 993 = (986 \times 1000) + 7^2 = 986,000 + 49$. On 4: I find this one easier using the $(a + b)^2$ formula from section 14.4: $(500 + 12)^2 = 500^2 + 2 \times 500 \times 12 + 12^2 = 250,000 + 12,000 + 144$. One could also slide apart by 12. On 5: The connection is to section 13.2. The number of combinations is $26 \times 26 \times 10 \times 10 \times 10 \times 10$. On 6: Students need to know the rules for exponents: $2^{20} = 2^{10} \times 2^{10}$. This is a good way to do the computation because $2^{10} = 1024$ is close to 1000 which makes the big slide method easier: $1024 \times 1024 = (1000 \times 1048) + 24^2 = 1,048,000 + 576$.

15.1: The Cube

1. Word Problem: Celeste wanted to make 5 cubes. How many marshmallows and toothpicks will she need?

   40 marshmallows and 60 toothpicks

2. A cube looks different from different perspectives. Using your cube model figure out what angle you would look at a solid cube from to make it look like each of the following outlines and draw in dashed lines to show where the sides are.

3. A triangular prism is the shape that you get when you make two identical triangles and then use three toothpicks to connect them together with one directly above the other. Make a triangular prism. How many vertices (marshmallows) and edges (toothpicks) does it have? How many triangular faces does it have? How many square faces does it have? How many faces does it have in total? What is its Euler characteristic (V − E + F)?

   V = _6_   E = _9_   F = _5_   V − E + F = _2_

   Triangular faces = _2_   Square faces = _3_

4. Making Connections: In chapter 10 we learned that there are 144 square inches in one square foot. By sharing cubes with the other kids in your class and stacking them (or making more on your own) figure out how many cubic feet there are in one cubic yard.

   27

5. Challenge: Use the following facts to try to figure out how much one gallon of water weighs: one cubic centimeter of water weighs one gram; one liter is one thousand cubic centimeters; a liter of water is about 33.8 fluid ounces; a gallon is four quarts; a quart is 32 fluid ounces; and a kilogram is about 2.2 pounds. (You can use a calculator if you want.) If you have a full gallon of water (or milk) and a scale where you are, check your calculation by weighing yourself while holding the gallon and subtracting your weight while not holding it.

   One gallon of water weighs about 8⅓ pounds

Notes on 1: Each cube needs 8 marshmallows and 12 toothpicks. On 5: It is probably easiest to use a few intermediate steps. One liter weighs 1000 grams = 1 kilogram. One gallon is 4 × 32 = 128 ounces = (128 ÷ 33.8) ≈ 3.8 liters. So a gallon weighs about 3.8 kilograms which is about 3.8 × 2.2 = 8.36 pounds. (The actual weight is closer to 8⅓ if you keep track of more decimal places.) Milk is a bit heavier than water. A gallon of whole milk weighs about 8.6 pounds plus the weight of the container. Skim milk is closer to 8.4. Temperature also matters – a cold gallon is about 0.02 pounds more than a warm one.

Name _____ **Answer Key** _____

15.2: The Tetrahedron

1. Word Problem: Zoe's teacher gave her 20 marshmallows and 25 toothpicks. How many complete tetrahedra can Zoe make?

2. Figure out how to hold a tetrahedron to make it look like each of the outlines below and draw in lines to show where the edges are.

   The second one can have a vertical line instead of a horizontal line.

3. Make three slightly shorter toothpicks by breaking off the end of a toothpick slightly less than one-third of the way along. Then build a tetrahedron that is flatter than a regular tetrahedron by making a triangle out of regular-sized toothpicks and then using three shorter toothpicks to connect the triangle to a marshmallow above the triangle. Place your short tetrahedron so that it is sitting on one of its smaller faces and look at another of the smaller faces from the side. What do you notice about the angles?

   The three smaller faces are right triangles. This makes one edge vertical.

4. Challenge: Get together in groups and make twelve more slightly shorter toothpicks as above. Build a regular tetrahedron will full-length toothpicks. Then use the shorter toothpicks to build short tetrahedrons coming out of each of the faces your regular tetrahedron. (Think of each face of the regular tetrahedron as a triangle and just add the three short toothpicks and one new marshmallow.) What shape do you have when you're done?

   A cube

   Notes on 1: Each tetrahedron needs 4 marshmallows and 6 toothpicks. After making four tetrahedrons you have 4 marshmallows and one toothpick left over, but that's not enough to make another.

15.3: Nets

1. Which of the following nets can be folded up to make a cube?

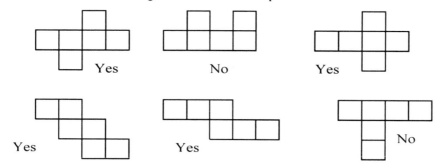

         Yes   No   Yes

Yes   Yes   No

2. Which of the following nets can be folded up to make a tetrahedron?

     Yes   No   Yes

     Yes

3. In each of the nets below color in the square that will be opposite the shaded square when the net is folded up to make a cube. Check your answers by drawing nets like these on paper, cutting them out, and folding them up.

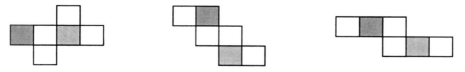

4. Making Connections: What shape would the net below make if it was folded up?

> A tetrahedron with three right triangles like the one in problem 3 of worksheet 15.2.

5. Challenge: Mei wanted to make as large of a cube as possible by drawing a net on an 8.5 inch by 11 inch piece of paper, cutting out the net, and then folding it up to make a cube. What is the length of the edge of the largest cube she can make?

> 2¾ inches

Notes on 1 and 2: A good way to do these is to pick out some square/triangle in the middle to think about as the bottom, and then to envision where each face goes as you fold up the shape. The ones that don't work end up with some faces doubly covered and others empty. On 5: You can make a cube that is 2¾ inches on a side by using the first, third, or fourth nets from question 1 with the sides of the shapes parallel to the sides of the paper. One could imagine that you could do better putting some net diagonally, but it turns out to not work better.

Name _____**Answer Key**_____

15.4: What Regular Polyhedra Could Exist?

1. Word Problem: Professor Ellison didn't want students to have to spend too long on the chapter 15 worksheets – the main idea of chapter 15 is to let students build things with marshmallows and toothpicks. How many questions did he decide to ask on the section 15.4 worksheet?

2. Suppose you cut the edges of a paper tetrahedron and fold down the three sides. What is the sum of the degree measures of the three angles that are adjacent to the vertex that the arrow points to?

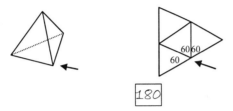

180

3. In the net below what is the addition problem describing the sum of the degree measures of the angles that meet at the vertex that the arrow points to?

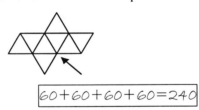

60+60+60+60=240

4. In a square pyramid the shapes that meet at each vertex are not identical. How many different-looking vertexes are there and what are the angles of the shapes that meet at each of them? Do the angles that meet at each vertex add up to less than 360 degrees?

There are two different vertices.
Four 60° angles meet at the top vertex.
One 90° angle and two 60° angles meet at each of the four bottom vertexes.
Each sum is less than 360°.

15.5: Building the Octahedron, Dodecahedron, and Icosahedron

1. How many marshmallows and toothpicks were used in making an octahedron? How many triangular faces does it have? What is V – E + F?

   V = _6_   E = _12_   Faces = _8_   V – E + F = _2_

2. When viewed from many angles the outline of an octahedron looks like a square. But it looks different from some angles. Figure out what angle to look at your octahedron from to make it look like each of the outlines below and draw in where the edges are.

3. Hold your icosahedron so that one marshmallow is on the very top and another is at the bottom resting on the table. If you look at the icosahedron from the side you'll notice that the marshmallows are in distinct layers defined by how far they are above the table. How many marshmallows are in each layer? How many marshmallows are there in all?

   > One at the top. Then two layers of 5. Then one on the bottom. 12 in all.

4. Ms. Munkres gave each student in her class 35 marshmallows and 50 toothpicks. Eva, however, ate 10 of the marshmallows. How many icosahedra can she now make?

5. What shape will the net below make if it is folded up? If you have time (and access to a photocopy machine) make a bigger copy of it and see if you can fold it up and tape it together.

   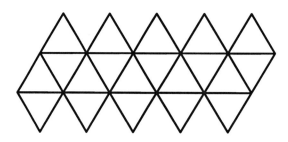

   An icosahedron

Notes on 4: Each icosahedon needs 30 toothpicks. (There are 20 faces and each has 3 edges, but this is double counting because each edge belongs to two faces.) It only requires 12 marshmallows, so the fact that Eva ate 10 isn't a problem. On 5: The fact that the net involves 20 triangles is a giveaway that it couldn't be any regular polyhedron other than an icosahedron. It will work better if you copy it on thick paper and do preliminary neat folds along each line before you try to make it an icosahedron.

Name _____**Answer Key**_____

15.6: Cutting Corners: The Soccer Ball

1. How many vertices, edges, and faces does a cuboctahedron have?

    Check you work by filling in the table below. In the final step you divide the number of edges by two because each edge belongs to two faces, and the number of vertices by four because each vertex belongs to four faces.

Type of Face	Number of Faces	Number of Edges	Number of Vertices
Square	6	_6_ faces × 4 = _24_	_6_ faces × 4 = _24_
Triangle	8	_8_ faces × 3 = _24_	_8_ faces × 3 = _24_
		Sum: 48	Sum: 48
	Total: 14	Divided by 2 → 24	Divided by 4 → 12

2. A cuboctahedron is made by cutting off the corners of a cube half way along each edge. Suppose that you instead cut off smaller corner pieces from each corner as in the picture below. How many faces would the resulting shape have? What shapes would they be? Make such a shape with Play-Doh to check your answer.

    14 faces. 6 octagons and 8 triangles.

3. How many vertices, edges, and faces does a soccer ball have? (A good way to check your work is to see if your answers have V − E + F = 2.

    60 vertices, 90 edges, and 32 faces.

4. Challenge: Make six extra-strong double-length edges by connecting doubled toothpicks end-to-end with another pair of toothpicks using a single marshmallow. Then use your six edges to make a double-size tetrahedron. Use single toothpicks – use a different color if you have toothpicks in multiple colors – to connect the marshmallows that are in the middle of each of the long sides with the other marshmallows that you can reach with a single toothpick. If you look at the single toothpicks you'll notice that they make a polyhedron. Does it look familiar?

    It is an octahedron

    Notes on 3: Think of making a soccer ball by cutting off each of the 12 vertices of an icosahedron one-third of the way along each edge turning the 20 triangles into 20 hexagons and creating 12 pentagons. 12+20 = 32 faces. The number of edges is (12 × 5 + 20 × 6) ÷ 2. The number of vertices is (12 × 5 + 20 × 6) ÷ 3 (because each vertex belongs to three faces.) On 4: A tipoff that the polyhedron will be an octahedron is that is has 6 vertices (the centers of the six edges of the tetrahedron), 12 edges (3 on each face of the tetrahedron), and triangular faces.

15.7: Using the Euler Characteristic to Figure Out Regular Polyhedra

1. Suppose that you want to make a regular polyhedron with three squares meeting at each vertex. If there are F squares then the number of edges will be $E = 4 \times F \div 2$ because each square has 4 edges, but adding them up the edges of each face is double counting. The number of vertices would be $V = 4 \times F \div 3$. Use these relations to figure out how many faces the polyhedron must have. ☐ 6

Guess for # of faces	# of Vertices ($4 \times F \div 3$)	# of Edges ($4 \times F \div 2$)	# of Faces	$V - E + F$
4	5⅓	8	4	1⅓
5	6⅔	10	5	1⅔
6	8	12	6	2
7	9⅓	14	7	2⅓

2. Fill in the table below to figure out how many faces a polyhedron must have if it has three pentagons meeting at each vertex. What polyhedron is this? ☐ A dodecahedron

Guess for # of faces	# of Vertices ($5 \times F \div 3$)	# of Edges ($5 \times F \div 2$)	# of Faces	$V - E + F$
3	5	7½	3	½
6	10	15	6	1
9	15	22½	9	1½
12	20	30	12	2
15	25	37½	15	2½

3. Making Connections: If a polyhedron has five triangles meeting at each vertex, then the number of edges will be $E = 3 \times F \div 2$ and the number of vertices will be $V = 3 \times F \div 5$. The fact that E and V are both whole numbers implies that two different prime numbers must be factors of F. What are they?

☐ 2 and 5

4. Challenge: Suppose a polyhedron has both hexagonal and triangular faces with two hexagons and one triangle meeting at each vertex. If there are V vertices then the number of triangular faces will be $V \div 3$ and the number of hexagonal faces will be $2 \times V \div 6$. Using these facts can you figure out how many faces of each type the polyhedron will have? (Hint: Make a table of possible guesses for V.)

☐ 4 faces will be hexagons and 4 will be triangles.

Notes on 3: The connection is to chapter 7. $3 \times F \div 2$ is a whole number only if F is a multiple of 2. And $3 \times V \div 5$ is a whole number only if V is a multiple of 5: On 4: A good way to do the problem is to make a table like those in problems 1 but using a guess for the # of vertices in the first column. Note that V must be a multiple of 3 to make $V \div 3$ a whole number. So the rows can be V = 3, 6, 9, …. The number of edges is $V \times 3 \div 2$. V = 12 turns out to make $V - E + F = 2$. You can make the shape by cutting off the corners of a tetrahedron one third of the way along each edge.

94

16.1: The Prime Factorization Number System

1. Word Problem: Melody wrote 55 as a product of two prime numbers. What were the numbers?

   $\boxed{5 \text{ and } 11}$

2. Use the following products to help figure out how to write each of the numbers below in the prime factorization number system.

   $300 = 3 \times 2 \times 5 \times 2 \times 5 = 2^2\,3^1\,5^2$

   $12 = 2 \times 2 \times 3 = 2^2\,3^1$

   $35 = 5 \times 7 = 5^1\,7^1$

   $216 = 2 \times 3 \times 2 \times 3 \times 2 \times 3 = 2^3\,3^3$

   $441 = 3 \times 3 \times 7 \times 7 = 3^2\,7^2$

   $1000 = 2 \times 5 \times 2 \times 5 \times 2 \times 5 = 2^3\,5^3$

3. Making Connections: Joe tried to write some numbers in the prime factorization number system but he made mistakes in three of the four problems that he did. Figure out which of the expressions below are wrong and explain how you know this from basic divisibility rules.

   $511 = 3^2\,57^1$ *Wrong! 511 isn't a multiple of 9 because 5+1+1 = 7. Also, 57 isn't prime: it's a multiple of 3.*

   $184 = 3^1\,61^1$ *Wrong! 184 is even. Also, it is not a multiple of 3 because 1+8+4=13).*

   $255 = 255^1$ *Wrong! 255 is a multiple of 5 (and again of 3 also).*

   $308 = 2^2\,7^1\,11^1$ *Correct.*

4. Challenge: How would you write the number 289,000 in the prime factorization number system?

   $\boxed{2^3\,5^3\,17^2}$

Notes on 3: The connection is to section 11.4. A number is multiple of 9 if the sum of the digits is a multiple of 9. It is a multiple of 3 if the sum of the digits is a multiple of 3. It is a multiple of 2 if the last digit is even. It is a multiple of 5 if the last digit is 0 or 5. On 4: A good first step is to recognize that $289{,}000 = 289 \times 1000$. Using $1000 = 10 \times 10 \times 10$ gives the $2^3 5^3$ part. Students may remember that $289 = 17^2$ from section 14.1. If not, they can recognize that it's not a multiple of 2, 3, or 5 and then try to find it's factors by dividing by 7, 11, 13, …

16.2: Unmultiplying aka Factor Trees

1. Fill in the missing numbers in the factor trees below

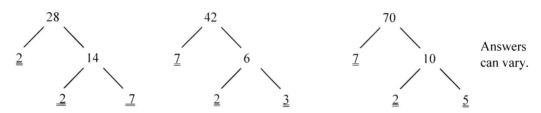

2. Make factor trees for 30, 148, and 308.

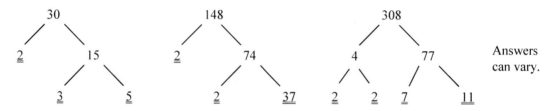

3. Draw factor trees (or do something else) to figure out how to write 40, 54, 108, and 504 in the prime factorization number system.

    $40 = 2^3\ 5^1$

    $54 = 2^1\ 3^3$

    $108 = 2^2\ 3^3$

    $504 = 2^3\ 3^2\ 7^1$

4. Making Connections: The number 9,996 can be written as a difference of squares: $9996 = 100^2 - 2^2$. Use the difference-of-squares formula to help figure out how to write 9996 in the prime factorization number system.

    $$9996 = 2^2\ 3^1\ 7^2\ 17^1$$

5. Challenge: Prime numbers have factor trees with just one level. Some numbers like 15 have very simple factor trees with just two levels. What is the smallest number whose factor tree cannot be written on just two levels? What is the smallest number whose factor tree cannot be written on three levels?

    8 and 32

    Notes on 2: The factor trees can be drawn in different ways. Many will find it more natural to start with 2×154 when doing 308. On 4: The connection is to section 9.4. The difference of squares formula gives $100^2 - 2^2 = (100 - 2) \times (100 + 2)$. A good second step is to factor out the 2's: $98 = 2 \times 49$ and $102 = 2 \times 51$. On 5: A number cannot be written on two levels if there are three primes (distinct or not) at the bottom of the factor tree. 2^3 is the smallest number with three prime factors. It cannot be written on two levels if it has 5 numbers on the bottom, and 2^5 is the smallest number with at least five prime factors.

Name _____**Answer Key**_____

16.3: What is the Prime Factorization Number System Good For?

1. Compute each of the products below. Write your answer in the prime factorization number system.

   $3^2\,5^1 \times 3^2 = 3^4\,5^1$ $2^2\,3^1 \times 3^2 = 2^2\,3^3$

   $3^2\,5^1 \times 3^1\,5^3 = 3^3\,5^4$ $2^2\,3^1 \times 2^2\,5^2 = 2^4\,3^1\,5^2$

   $2^2\,17^1 \times 2^5\,5^1\,17^2 = 2^7\,5^1\,17^3$ $3^2\,13^1 \times 2^2\,5^2\,13^{12} = 2^2\,3^2\,5^2\,13^{13}$

2. Circle the numbers below that are perfect squares.

   $2^1\,3^3\,5^5$ $\boxed{2^2\,3^4\,5^{10}7^6}$ $\boxed{13^2\,17^4}$ $2^2\,7^4\,13^2\,23^1$

3. Find the square roots of the following numbers.

   $\sqrt{2^2 3^4 5^2} = 2^1\,3^2\,5^1$ $\sqrt{2^6 3^4} = 2^3\,3^2$ $\sqrt{2^6 5^4 17^2} = 2^3\,5^2\,17^1$

4. Making Connections: How many different ways are there to order the letters ABCDEF? Write your answer in the prime factorization number system.

   $\boxed{2^4\,3^2\,5^1}$

5. Challenge: What is the cube of 54? Write your answer in the prime factorization number system.

   $\boxed{2^3\,3^9}$

Notes on 2: A number is a perfect square if all exponents are even. On 4: The connection is to section 13.3. The number of ways to order the six letters is $6 \times 5 \times 4 \times 3 \times 2 \times 1$ because there are 6 ways to choose the first letter, then 5 remaining choices for the second letter, etc. To write this in the prime factorization number system don't multiply this out: instead just put each number in the product in the prime factorization number system, e.g. $6 = 2^1\,3^1$ and $4 = 2^2$, and add the exponents. On 5: Don't start by cubing 54 in base 10. Instead, start by writing $54 = 2^1\,3^3$ and then cube this by multiplying each exponent by 3.

16.4: What is the Prime Factorization Number System Not Good For?

1. Word Problem: Wei wants to buy an iPod touch. Best Buy is selling it for $\$2^1 3^4$. Target is selling it for $\$3^1 5^1 11^1$. Where should she buy it?

2. Circle the larger number in each of the pair of numbers below.

   $(3^2 5^1)$ or 3^2 $3^2 5^1$ or $(3^2 7^1)$

   $2^2 5^1$ or $(2^2 5^2)$ $(2^{11} 7^2)$ or $2^{11} 5^2$

   $(3^2 5^1 7^2)$ or $3^2 5^1 7^1$

3. Find each of the following sums. Write your answers in the prime factorization number system. (Using the distributive property makes them easier.)

   $2^1 3^3 5^1 + 2^1 3^3 = 2^1 3^3 (5 + 1) = 2^1 3^3 (2^1 3^1) = 2^2 3^4$

   $2^1 3^2 + 2^1 5^1 = 2^1 (9 + 5) = 2^1 (2^1 7^1) = 2^2 7^1$

   $2^2 3^1 + 2^2 5^1 = 2^2 (3 + 5) = 2^2 (2^3) = 2^5$

   $7^1 11^3 + 7^1 11^2 = 7^1 11^2 (11 + 1) = 7^1 11^2 (2^2 3^1) = 2^2 3^1 7^1 11^2$

4. Making Connections: What is $3^2 17^2$ in the base 10 number system?

   2601

5. Challenge: What is the sum of the first 20 multiples of 3? Write your answer in the prime factorization number system.

   $2^1 3^2 5^1 7^1$

   Notes on 1: The prices at the two stores are $162 and $165. On 4: The connection is to section 14.2 (or 14.4). $3^2 17^2$ is the square of $3^1 17^1 = 51$. The slide method says this can be computed as $50 \times 52 + 1$. The $(a + b)^2$ method says it is equal to $50^2 + 2 \times 50 + 1$. On 5: Using the distributive property (section 9.1) the sum is $3 \times (1 + 2 + \ldots + 20)$. Adding pairs of terms from the outside in (as in section 3.3) we see that $1 + 2 + \ldots + 20 = 10 \times 21$. So the answer is $3^1 \times 2^1 5^1 \times 3^1 7^1 = 2^1 3^2 5^1 7^1$.

16.5: A Shortcut for Some Multiplications: Unmultiply, Reorder, and Remultiply

1. For each of the products below write down an ordering that would make the multiplication easy to compute.

    $2 \times 2 \times 3 \times 5 \times 5 = (2 \times 5) \times (2 \times 5) \times 3$

    $2 \times 2 \times 5 = 2 \times (2 \times 5)$

    $5 \times 47 \times 2 = (2 \times 5) \times 47$

    $3 \times 3 \times 7 \times 11 = (3 \times 3 \times 11) \times 7$

2. Figure out each of the products below by unmultiplying, reordering, and remultiplying.

    $28 \times 25 = 2^2 \ 7^1 \times 5^2 = (2 \times 5) \times (2 \times 5) \times 7 = 700$

    $8 \times 25 = 2^3 \times 5^2 = (2 \times 5) \times (2 \times 5) \times 2 = 200$

    $48 \times 625 = 2^4 \ 3^1 \times 5^4 = (2 \times 5) \times (2 \times 5) \times (2 \times 5) \times (2 \times 5) \times 3 = 30,000$

3. Use the products 9×11=99, 7×11×13=1001, or 3×37=111 to help figure out the products below.

    $55 \times 91 = 5 \times 11 \times 7 \times 13 = 5 \times (7 \times 11 \times 13) = 5 \times 1001 = 5005$

    $3 \times 74 = 3 \times 2 \times 37 = (3 \times 37) \times 2 = 111 \times 2 = 222$

    $18 \times 33 = 2 \times 3 \times 3 \times 3 \times 11 = (2 \times 3) \times (3 \times 3 \times 11) = 6 \times 99 = 594$

4. Making Connections: What is the area in square inches of a triangle with a base of 75 inches and a height of 16 inches?

    600 square inches

5. Challenge: What is 256×125^3?

    500,000,000

Notes on 4: The connection is to section 10.4. The area of a triangle is ½ base × height. In this problem you could easily use long multiplication. Or using the trick of this section it is ½ × 75 × 16 = 75 × 8 = (3 × 25) × (2 × 4) = 3 × 2 × (25 × 4) = 600. On 5: $256 = 2^8$. $125 = 5 \times 5 \times 5$ so $125^3 = (5 \times 5 \times 5) \times (5 \times 5 \times 5) \times (5 \times 5 \times 5) = 5^9$. So $256 \times 125^3 = 2^8 \times (5 \times 5^8)$ $= 5 \times 10^8 = 500,000,000$.

16.6: Which Numbers are Factors of 102?

1. Word Problem: There are 21 students in Ms. Hoover's class. She wants to divide the class into teams with an equal number of students for a game. How many teams could she make?

   3 or 7 (also 21 if you allow teams of one)

2. In each of the rows below circle the numbers that are factors of the first number and cross out the ones that are not.

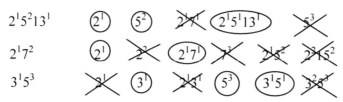

   $2^1 5^2 13^1$ (2^1)   (5^2) $2^1 5^1$ ($2^1 5^1 13^1$) 5^3

   $2^1 7^2$ (2^1) 2^2 ($2^1 7^1$) $2^1 5^1$ $2^1 3^2$ $2^3 5^2$

   $3^1 5^3$ 3^2 (3^1) $3^2 5^1$ (5^3)   ($3^1 5^1$) $3^2 5^3$

3. Find the prime factorization of the first number in each row and use this to help figure out whether the other numbers are factors of the first number. Circle the ones that are.

   15   ① ③ ⑤ 10 ⑮ 25

   28   ② ⑦ 12 ⑭ 21 ㉘

   63   ① ③ 5 ⑦ ⑨ ㉑

4. List all of the factors of each number. Write the factors in the prime factorization number system.

   30   1 2^1 3^1 5^1 $2^1 3^1$ $2^1 5^1$ $3^1 5^1$ $2^1 3^1 5^1$

   10   1 2^1 5^1 $2^1 5^1$

   70   1 2^1 5^1 7^1 $2^1 5^1$ $2^1 7^1$ $5^1 7^1$ $2^1 5^1 7^1$

5. Making Connections: Use the difference of squares formula to help figure out all of the factors of 399.

   1, 3, 7, 19, 21, 57, 133, 399

6. Challenge: Find all of the factors of 2002.

   1, 2, 7, 11, 13, 14, 22, 26, 77, 91, 143, 154, 182, 286, 1001, 2002

   Notes on 5: The connection is to section 9.4. To find the prime factorization of 399 we can write $399 = 400 - 1$ and then use the difference of squares formula: $399 = 20^2 - 1^2 = (20 - 1) \times (20 + 1) = 19 \times (3 \times 7)$. The prime factors are then 1, 3, 7, 19, $3^1 7^1$, $3^1 19^1$, $7^1 19^1$, and $3^1 7^1 19^1$. On 6: The obvious first step for making a factor tree is $2002 = 2 \times 1001$. Students can then either remember that $1001 = 7 \times 11 \times 13$ from section 16.5 or find this by testing for divisibility by primes bigger than 5. Given that $2002 = 2 \times 7 \times 11 \times 13$, the factors of 2002 are 1, 2, 7, 11, 13, $2^1 7^1$, $2^1 11^1$, $2^1 13^1$, $7^1 11^1$, $7^1 13^1$, $11^1 13^1$, $3^1 7^1 11^1$, $3^1 7^1 13^1$, $3^1 11^1 13^1$, $7^1 11^1 13^1$, and $3^1 7^1 11^1 13^1$.

Name _____**Answer Key**_____

16.7: The Prime Monster Number System

1. Suppose that you had been asked to draw the pages for the 15 and 28 monsters for Prof. Schwartz's book *You Can Count on Monsters*. What monsters would appear in symbol for 15? What monsters would appear in the symbol for 28? Make up symbols for 15 and 28 using his 2, 3, 5, and 7 monsters.

   The symbol for 15 would have a 3 monster and a 5 monster.
   The symbol for 28 would have two 2 monsters and a 7 monster.

2. Suppose you were going to make up your own version of the prime monster number system. Draw pictures of the monsters you'd make up for the numbers 2 and 3. What kinds of personalities would they have?

3. Make up a symbol for some bigger prime p in your prime monster number system and then make up a symbol for 2^3p^1.

17.1: Equivalent Fractions: One Fraction, Many Names

1. Word Problem: Anna can eat two-thirds of a small Bertucci's pizza. How many slices is this if a small pizza has 6 slices?

   4 slices

2. Convert the following fractions to twenty-fourths.

   $$\frac{1}{4} = \frac{6}{24} \qquad \frac{1}{3} = \frac{8}{24} \qquad \frac{5}{8} = \frac{15}{24} \qquad \frac{3}{4} = \frac{18}{24}$$

3. Fill in the missing numerators or denominators to make equivalent fractions.

   $$\frac{1}{3} = \frac{6}{18} \qquad \frac{2}{3} = \frac{12}{18} \qquad \frac{5}{9} = \frac{10}{18} \qquad \frac{3}{8} = \frac{33}{88}$$

   $$\frac{1}{2} = \frac{81}{162} \qquad \frac{3}{10} = \frac{111}{370} \qquad \frac{1}{5} = \frac{199}{995} \qquad \frac{2}{3} = \frac{446}{669}$$

4. Making Connections: For what value of x are $\dfrac{1}{85}$ and $\dfrac{85}{x}$ equivalent fractions?

   7225

5. Challenge: How many fractions with a two-digit odd denominator are equivalent to one-third?

   15

Notes on 4: The connection is to section 14.2. The answer is 85^2. The sliding apart method imples that this can be computed easily as $80 \times 90 + 5^2 = 7200 + 25 = 7225$. On 5: The denominator must be a two digit odd number that is a multiple of 3. One could count them by trying all odd numbers and finding that 15, 21, 27 ... 99 work. But it's quicker to recognize the pattern: the numbers that work will be the product of 3 and an odd number so the numbers that work are 3×5, 3×7, 3×9, ..., 3×33. So the number of possible denominators is the number of elements in the set {5, 7, 9, ..., 33}. There are 15 such numbers.

17.2: Which Fraction is Bigger?

1. Word Problem: Noemi likes cake but doesn't eat much so she always prefers the smaller piece when given a choice? Which would she prefer: one-fifth of a cake or two-elevenths?

$\boxed{\text{Two elevenths}}$

2. Write >, =, or < to compare each pair of fractions.

$$\frac{1}{12} < \frac{2}{12} \qquad \frac{3}{13} < \frac{5}{13} \qquad \frac{5}{11} > \frac{5}{13} \qquad \frac{2}{111} > \frac{2}{131}$$

3. Use cross-multiplication to figure out which fraction in each pair is bigger.

$$\frac{1}{3} > \frac{11}{35} \qquad \frac{2}{7} > \frac{5}{18} \qquad \frac{1}{10} < \frac{3}{28} \qquad \frac{6}{117} < \frac{17}{148}$$

4. Making Connections: Which is larger: $\dfrac{29}{98}$ or $\dfrac{33}{102}$? (Hint: Think about what happens when you move numbers that you're multiplying apart or use the distributive property.)

$$\boxed{33/102}$$

5. Challenge: The Fibonacci numbers are 1, 1, 2, 3, 5, 8, 13, 21, 34, … (Each number is the sum of the two earlier numbers.) Can you find a pattern to figure out which fraction is bigger when we compare fractions made out of adjacent Fibonacci numbers? For example, 'which is bigger 2/3 or 3/5?' and 'which is bigger 3/5 or 5/8?'

In the sequence 1/1, 1/2, 2/3, 3/5, 5/8, …each odd term (e.g. 1/2, 2/3, 5/8) is bigger than the terms that come before or after.
Comparing within the odd terms the earlier terms are bigger.
Comparing within the even terms the earlier terms are smaller.

Notes on 4: The connection is to section 9.2 or 14.2. Using cross-multiplication the first fraction is bigger if and only if $29 \times 102 > 33 \times 98$. The first is $(30 - 1) \times (100 + 2) = 3000 - 100 + 60 - 2$. The second is $(30 + 3) \times (100 - 2) = 3000 + 300 - 60 - 6$. The second is bigger. Alternately, note that $29 + 102 = 33 + 98$ so you can think of the first pair of numbers as numbers that have been shifted farther apart. As in section 14.2 the product of two numbers gets smaller as they are shifted farther apart. On 5: A good way to see the patterns is with funny subtraction. For example, applying funny subtraction to the third and fourth terms we see that 2/3 > 3/5 if and only if 2/3 > 1/2, i.e. the third term is bigger than the fourth if the second is less than the third. Repeating this comparison shows that the odd terms are always bigger than the surrounding even terms. Comparing consecutive odd terms 5/8 is bigger than 2/3 if and only if 3/5 is bigger than 2/3. It is not (3/5 is an even term) so 5/8 is smaller. This explains why the odd terms get smaller as you go along.

HMES © 2013 Glenn Ellison

Name _____ **Answer Key** _____

17.3: Multiplying Fractions

1. Word Problem: Tian brought one-third of a cake home after a party at her office. Kevin ate two-fifths of it. What fraction of the whole cake did he eat?

$$2/15$$

2. Draw in more lines and shade in parts of each rectangle to show that 2/3 of 1/6 is 2/18 and 3/4 of 2/5 is 6/20.

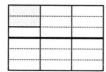

3. Give the answer to each of the following multiplication problems?

$$\frac{1}{4} \times \frac{1}{6} = \frac{1}{24} \qquad \frac{2}{5} \times \frac{7}{13} = \frac{14}{65} \qquad \frac{1}{10} \times \frac{3}{28} = \frac{3}{280} \qquad \frac{1}{101} \times \frac{23}{77} = \frac{23}{7777}$$

4. What is one-third of five-sevenths?

$$5/21$$

5. Making Connections: In the base 8 number system what is $\dfrac{3_{(8)}}{10_{(8)}} \times \dfrac{3_{(8)}}{13_{(8)}}$?

$$\frac{11_{(8)}}{130_{(8)}}$$

6. Challenge: Catherine has invented a method for drawing snowflakes. She starts with an equilateral triangle that is one inch long on each side. She then erases the middle third of each of the sides and replaces it with two sides of the same length that stick out like an equilateral triangle. This leaves her with a 12-sided shape that looks like a six-pointed star. She then does the same thing again: she erases the middle third of each of the 12 sides and replaces each with two pieces of the same length that stick out like a smaller equilateral triangle. After which step is the perimeter of her figure first longer than 7 inches?

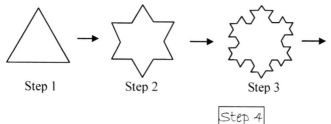

Step 1   Step 2   Step 3

$$\text{Step 4}$$

Notes on 4: The connection is to chapter 2. For the denominator it helps to know you can multiply a base 8 number by $10_{(8)}$ simply by adding a zero. At each stage each line segment is replaced by four segments that are each one-third of its length, so the perimeter is multiplied by 4/3. At step 4 the perimeter is $3 \times (4/3)^3 = 3 \times (64/27) = 64/9$.

Name _____**Answer Key**_____

17.4: Adding Fractions

1. Word Problem: Jack ate one-third of a large pizza. Luke ate one-quarter of the pizza. What fraction of the pizza was left for their parents?

$$\boxed{5/12}$$

2. Add each of the pairs of fractions.

$$\frac{1}{13}+\frac{1}{13}=\frac{2}{13} \qquad \frac{5}{11}+\frac{3}{11}=\frac{8}{11} \qquad \frac{1}{12}+\frac{5}{12}=\frac{6}{12}=\frac{1}{2} \qquad \frac{13}{102}+\frac{4}{102}=\frac{17}{102}=\frac{1}{6}$$

3. Add each pair of fractions by putting them over a common denominator.

$$\frac{1}{2}+\frac{1}{5}=\frac{7}{10} \qquad \frac{1}{3}+\frac{1}{8}=\frac{11}{24} \qquad \frac{3}{7}+\frac{2}{3}=\frac{23}{21} \qquad \frac{5}{14}+\frac{2}{3}=\frac{43}{42}$$

$$\frac{1}{2}+\frac{1}{3}=\frac{5}{6} \qquad \frac{2}{3}+\frac{1}{7}=\frac{17}{21} \qquad \frac{1}{10}+\frac{2}{9}=\frac{29}{90} \qquad \frac{2}{5}+\frac{2}{11}=\frac{32}{55}$$

4. Add each pair of fractions by putting them over a common denominator. (The problems are easier if you use a denominator that's smaller than the product of the two denominators.)

$$\frac{1}{2}+\frac{3}{8}=\frac{7}{8} \qquad \frac{2}{11}+\frac{1}{55}=\frac{1}{5} \qquad \frac{4}{25}+\frac{3}{10}=\frac{23}{50} \qquad \frac{5}{21}+\frac{3}{35}=\frac{34}{105}$$

5. Making Connections: What is $\dfrac{30}{59}+\dfrac{30}{61}$?

$$\frac{3600}{3599}$$

6. Challenge: What is $\dfrac{2}{5}+\dfrac{3}{7}+\dfrac{4}{9}+\dfrac{5}{11}+\dfrac{8}{55}+\dfrac{8}{63}$?

____

Notes on 5: The connection is to sections 9.1 and 9.4. The numerator is $30 \times 59 + 30 \times 61$ $= 30 \times (59 + 61) = 30 \times 120 = 3600$. The denominator is $59 \times 61 = (60 - 1) \times (60 + 1) =$ $60^2 - 1 = 3600 - 1 = 3599$. The best way to add these fractions is to change the order. First, $\dfrac{2}{5}+\dfrac{5}{11}+\dfrac{8}{55}=\dfrac{22}{55}+\dfrac{25}{55}+\dfrac{8}{55}=\dfrac{55}{55}=1$. Putting the other fractions over a common denominator of 63 shows that they also add up 1.

HMES © 2013 Glenn Ellison

18.1 Basic Probability: Rolling One Die

1. Word Problem: Chairman Mao has bet a friend of his that a fair 6-sided die will come up as a 1. What is the probability that Mao wins the bet?

    1/6

2. Circle the numbers that satisfy the stated conditions:

   Equal to 3:   1   2 ③ 4   5   6

   Is prime: 1 ②③ 4 ⑤ 6

   Is odd: ① 2 ③ 4 ⑤ 6 ⑦ 8

   Is a multiple of 3:   1   2 ③ 4   5 ⑥ 7   8

   Is a perfect square:: ① 2   3 ④ 5   6   7   8

3. What is the probability of

   Rolling an even number on a fair 6-sided die   1/2

   Rolling a number less than 3 on a fair 6-sided die   1/3

   Rolling a multiple of five on a fair 8-sided die   1/8

   Rolling a prime on a fair 4-sided die   1/2

   Rolling a perfect square on a fair 6-sided die   1/3

4. Making Connections: What is the probability of rolling a composite number on a fair 12-sided die?

    1/2

5. Challenge: What is the largest value of N for which the probability of rolling a prime number on a fair N-sided die is exactly one-half?

   _____ 8 _____

   Notes on 4: The connection is to section 7.1 or 7.5. The composites are 4, 6, 8, 9, 10, and 12. The probability of rolling a prime on an 8-sided die is one-half because 2, 3, 5, and 7 are prime. The probability is less than one-half for all larger N because only four the first ten numbers are prime and then at most one of every two subsequent numbers is prime. (Every other number is even.)

Name _____ **Answer Key** _____

18.2 Problems with 2 Dice

1. Word Problem: Preksha rolls two four-sided dice. Draw a grid showing all of the combinations that could come up and find the probability that the sum of the two numbers is 7?

    1,1 1,2 1,3 1,4
    2,1 2,2 2,3 2,4
    3,1 3,2 3,3 (3,4)   → 2/16 = 1/8
    4,1 4,2 (4,3) 4,4

2. In the grid on the left circle all combinations that have a sum of 4. In the grid on the right circle all combinations in which the one number is exactly two more than the other.

    1,1 1,2 (1,3) 1,4 1,5 1,6   1,1 1,2 (1,3) 1,4 1,5 1,6
    2,1 (2,2) 2,3 2,4 2,5 2,6   2,1 2,2 2,3 (2,4) 2,5 2,6
    (3,1) 3,2 3,3 3,4 3,5 3,6   (3,1) 3,2 3,3 3,4 (3,5) 3,6
    4,1 4,2 4,3 4,4 4,5 4,6   4,1 (4,2) 4,3 4,4 4,5 (4,6)
    5,1 5,2 5,3 5,4 5,5 5,6   5,1 5,2 (5,3) 5,4 5,5 5,6
    6,1 6,2 6,3 6,4 6,5 6,6   6,1 6,2 6,3 (6,4) 6,5 6,6

3. Juliana rolls two six-sided dice. What is the probability that the sum of the two numbers is 4?

    1/12

4. After losing his previous bet, Mao bets his friend that, if the fair 6-sided die is rolled twice more, at least one time a 6 will be rolled. Find the probability that he wins this new bet.

    11/36

5. Making Connections: Jin rolls two six-sided dice. What is the probability that the product of the two numbers is a factor of 55?

    1/12

Notes on 5: The connection is to section 16.6. The factors of 55 are 1, 5, 11, and 55. The product of the numbers on two six-sided dice can never be 11 or 55. The combinations that give 1 or 5 are (1, 1), (5, 1), and (1, 5).

Name _____ __Answer Key__ _____

18.3 Random Lists or Groups

1. Word Problem: Anna, Bill, and Claire worked together on a sheet of math problems. Ms. Hoover randomly picks two of them to explain a solution at the board. If she is equally likely to pick any two of them, what is the probability that she picks the two girls?

1/3

2. Circle all of the combinations of three H's and T's below in which there are exactly two heads. If you flip a coin three times what is the probability that exactly two of the three flips will be Heads?

HHH (HHT)(HTH) HTT (THH) THT   TTH   TTT

3/8

3. If you flip a fair coin 3 times what is the probability that you get at least two Heads in a row at some point?

3/8

4. If you flip a fair coin 4 times what is the probability that you get exactly two Heads?

3/8

5. Anna puts scrabble tiles with the letters A, R, and T in a bag. She picks them out of the bag one at a time and puts them on the table in that order. What is the probability that she makes an English word by doing this?

1/2

6. Challenge: Kate puts tiles with the letters I, O, N, T, and W in a bag and draws three out at random. What is the probability that she can rearrange the letters she took out to form an English word?

7/10

Notes on 1: There are three possible combinations and one of them (Anna, Claire) is the two girls. On 3: The sequences with two heads in a row are HHH, HHT, and THH. On 4: The sequences with exactly two Heads are HHTT, HTHT, HTTH, THHT, THTH, and TTHH. On 5: There are 6 possible sequences: ART, ATR, RAT, RTA, TAR, TRA. Three are words: ART, RAT, and TAR. On 6: There are 10 possible three letter combinations: ION, IOT, IOW, INT, INW, ITW, ONT, ONW, OTW, and NTW. There are six possible orderings for each three-letter combinations. The ones that can be arranged in a word are: ION (ION), INT (TIN), INW (WIN), ITW (WIT), ONT (TON), ONW (OWN), and OTW (TOW). The ones that don't work are IOT, IOW, and NTW.

18.4 Complicated Problems: Don't Write It All Out, Just Count

1. Word Problem: The AMC 8 math contest is a 25 question multiple choice test: every answer is A, B, C, D, or E. The questions get harder as you go along so Balazs and Botond both guess randomly on number 25. What is the probability that they make the same guess?

    > 1/5

2. Find the following.

    o The probability that a 3-digit number, chosen randomly, begins with two 5's.

       > 1/90

    o The probability that the numbers rolled on a fair 20-sided die and a fair 8-sided die are the same.

       > 1/20

    o The probability that you get all Heads when you flip a fair coin four times.

       > 1/16

3. If you flip a fair coin 5 times what is the probability that exactly 4 of the flips come up heads?

    > 5/32

4. Making Connections: If you pick a random whole number from 1 to 100 with all numbers being equally likely, what is the probability that the number is a prime that has a 1 as its last digit?

    > 1/20

5. Challenge: What is the probability that a 4-digit number chosen at random is a multiple of 8?

    > 1/8

Notes on 1: There are 25 possible choices for the ordered pair of guesses. Five of them (A, A), (B, B), …, (E, E) have the two making the same guess. On 2: There are 90 possible two digit numbers: 10, 11, …, 99 and only 55 begins with two fives. The set of possible ordered pairs will be an 8 by 20 matrix and there are 8 combinations like (1, 1), …, (8, 8). There are 16 sequences for 4 coin flips. On 3: There are 32 sequences and 5 of them are like HHHHT. On 4: The connection is to section 7.5. The primes ending in 1 are 11, 31, 41, 61, and 71. On 5: The best way to do this is to think about listing the numbers 1000, 1001, …, 9999. You would circle 1000, then not circle the next 7 numbers, then circle 1008, then not circle the next 7 and so on. At the end of each block of 8 numbers you've circled exactly one-eighth of the numbers. And there are an even number of blocks of 8 so the answer is one-eighth. (You could also count that there are 9000 four digit numbers and 1125 of them (1000, … 9992) are multiples of 8, but this is harder.)

Name _____ **Answer Key** _____

18.5 How Do I Know If A Coin Is Fair?

1. Flip a coin ten times and count the number of Heads that come up. Put an X above the number. Then repeat this a few more times. Then ask the other kids how many Heads they got in each of their flips and combine everyone's X's into one histogram. Does it look like a bell-shaped curve?

    _____

    0 1 2 3 4 5 6 7 8 9 10

2. If you flip a fair coin N times then about 95% of the time the number of Heads that you get will be between $\frac{1}{2}N - \sqrt{N}$ and $\frac{1}{2}N + \sqrt{N}$. What are these two numbers if N=49? What are they if N=100?

    When N=49 they are 17½ and 31½. For 100 they are 40 and 60.

3. Practice spinning a penny by putting your left index finger on the top and flicking the right edge of the coin. Then spin your coin 10 times and count the number of Heads you get. Repeat this and record the number of Heads in four more sets of 10. Does the total number of Heads you got in 50 spins make you think that your penny probably isn't fair?

    Answers can vary, but most pennies aren't fair when you spin them.

4. Ask the other people in your class how many Heads they got in each of their sets of 10 spins. (If you're not doing this with a class do a few more sets of 10 yourself. Draw X's on the line below to show how many times Heads came up in each of the sets of 10 flips. Does it look like the histogram from question 1?

    _____

    0 1 2 3 4 5 6 7 8 9 10

    usually, this histogram looks different with many more big numbers.

5. Another alternative to flipping a coin is to balance a penny on its edge and then jiggle the table and see which side it falls onto. Balance 10 pennies on their edge on a table. Then shake the table and count how many end up Heads. Do this a few more times. Does it seem like Heads comes up half of the time?

    It has been reported that Heads comes up more than half the time.

© 2013 Glenn Ellison

19.1 Approximate Division

1. Word Problem: The Weasleys have a budget of 26 galleons (which is 442 sickles) for Gilderoy Lockhart books. Each book costs 39 sickles. How many books can they buy?

    11

2. Circle the correct answers for each of the following division problems:

    What integer is closest to $831 \div 8$?

    (a) 102   (b) 104   (c) 105   (d) 111   (e) 131   (b) 104

    What is $1001 \div 91$?

    (a) 9   (b) 11   (c) 13   (d) 33   (e) 91,091   (b) 11

    What is $1769 \div 29$?

    (a) 47   (b) 51   (c) 59   (d) 61   (e) 131   (d) 61

3. George got a total of 435 points on the five tests in his math class. What is his average score?

    87

4. A gallon is 128 ounces. How many one-gallon jugs of lemonade would you need to bring to a math meet if you wanted to have enough to give a 6 ounce cup to each of the 100 students who will be there?

    5

5. Making Connections: Jae's teacher brought 252 marshmallows and 17,187 toothpicks to class. How many toothpick-and-marshmallow octahedrons will the class be able to make?

    42

6. Challenge: The distance from the Earth to the Sun is about 150 million kilometers. Light travels at about 299,792 kilometers per second. About how many minutes and seconds does it take for light from the Sun to reach the Earth?

    About 8 minutes and 20 seconds

Notes on 1: $39 \times 11 = 390 + 39 = 429$. They have 13 sickles left over. On 2: A good first step on $1769 \div 29$ is to guess about 60 and use the distributive property: $29 \times 60 = (30 - 1) \times 60 = 1800 - 60 = 1740$. On 4: You need 600 ounces. $128 \times 5 = 640$. On 5: The connection is to section 15.5. They will obviously run out of marshmallows first. An octahedron needs 6 marshmallows. On 6: The questions asks you to find $150,000,000 \div 299,972$. If you round off 299,972 to 300,000 then 500 is an obvious first guess. $299,972 \times 500 = (300,000 - 28) \times 500 = 150,000,000 - 14,000$. The extra 14,000 miles will only take a small fraction of a second so 500 seconds is correct to the nearest second.

19.2 Dividing By One Digit Numbers

1. Word Problem: If 3 galleons are worth 1,479 knuts, how many knuts is one galleon?

   $\boxed{493}$

2. Use long division to find the answers to each of the problems below.

   $3\overline{)69}$ $\boxed{23}$ $3\overline{)243}$ $\boxed{81}$ $8\overline{)576}$ $\boxed{72}$ $5\overline{)375}$ $\boxed{75}$ $7\overline{)511}$ $\boxed{73}$

   $3\overline{)6,732}$ $\boxed{2,244}$ $2\overline{)2,532}$ $\boxed{1,266}$ $5\overline{)5,765}$ $\boxed{1,153}$ $9\overline{)3,762}$ $\boxed{418}$

3. Find the quotient and remainder in each of the division problems below.

   $3\overline{)539}$ $\boxed{179\ r2}$ $4\overline{)3,537}$ $\boxed{884\ r1}$ $7\overline{)52,763}$ $\boxed{7,537\ r4}$ $9\overline{)53,761}$ $\boxed{5,973\ r4}$

4. Making Connections: The measures of the interior angles of any dodecagon add up to 1800 degrees. If a dodecagon has three right angles and the other nine angles are all equal, what is the degree measure of the nine equal angles?

   $\boxed{170}$

5. Challenge: What is the remainder when 1,000,000,000,000,000 is divided by 7?

   _____ $\boxed{6}$ _____

   Notes on 4: The connection is to section 4.4 although the formula is also included in the question. The nine equal angles add up to $1800 - 3 \times 90 = 1530$. So the answer is $1530 \div 9 = 170$. On 5: Students can simply use long division which takes a while but works fine. In the process of doing this they may notice the repeating pattern in the quotients and remainders. The problem can also be done using modular arithmetic as in chapter 11. From that perspective, the problem asks you to simpify 10^{15} in mod 7 arithmetic. $10 \equiv 3 \bmod 7$ so you can figure this out by multiplying and observing the pattern: $3^1 \equiv 3$, $3^2 \equiv 3 \times 3 \equiv 2$, $3^3 \equiv 2 \times 3 \equiv 6$, $3^4 \equiv 6 \times 3 \equiv 4$, $3^5 \equiv 4 \times 3 \equiv 5$, $3^6 \equiv 5 \times 3 \equiv 1$, $3^7 \equiv 1 \times 3 \equiv 3$, and the pattern continues: 3, 2, 6, 4, 5, 1, 3, 2, 6, 4, 5, 1, ... The 15th term in this sequence is the same as the 3rd which is 6. You could also use that $3^{15} = (3^5)^3$.

19.3 Dividing By Bigger Numbers

1. Word Problem: A Boeing 757-200 has 224 seats. If you can put two penguins in each seat, how many planes would you need to charter to fly 7616 penguins from Boston to Greenland?

   17

2. Use long division to find the answers to each of the problems below.

$$\overset{\boxed{11}}{13\overline{)143}} \qquad \overset{\boxed{15}}{23\overline{)345}} \qquad \overset{\boxed{31}}{28\overline{)868}} \qquad \overset{\boxed{25}}{25\overline{)625}}$$

$$\overset{\boxed{148}}{113\overline{)16,724}} \qquad \overset{\boxed{515}}{102\overline{)52,530}} \qquad \overset{\boxed{1,515}}{215\overline{)325,725}}$$

3. Find the quotient and remainder in each of the division problems below.

$$\overset{\boxed{385\ r4}}{14\overline{)5,394}} \qquad \overset{\boxed{323\ r21}}{32\overline{)10,357}} \qquad \overset{\boxed{2085\ r\ 158}}{313\overline{)652,763}}$$

4. Making Connections: Kate's bedroom has an area of 34,560 square inches. How many square feet is this?

   $\boxed{\text{240 square feet}}$

Notes on 1: With two penguins per seat you can put 448 penguins on each plane. So the answer is 7616 ÷ 448 = 17. Or you could figure that it would take 7616 ÷ 224 = 34 planes if you were doing one per seat and then divide 34 by 2. On 4: The connection is to section 10.2. One square foot is 144 square inches so the answer is 34,560 ÷ 144 = 240.

19.3 Dividing By Bigger Numbers

5. Use long division to find the answers to each of the problems below.

    $\boxed{145}$ $\boxed{612}$ $\boxed{241}$
    $68\overline{)9,860}$ $245\overline{)149,940}$ $683\overline{)164,603}$

6. Find the quotient and remainder in each of the division problems below.

    $\boxed{72\ r66}$ $\boxed{191\ r43}$ $\boxed{3,062\ r169}$
    $74\overline{)5,394}$ $54\overline{)10,357}$ $187\overline{)572,763}$

7. Making Connections: Use the distributive property when multiplying to make it easier to use long division to find the answers to the problems below.

    $\boxed{343}$ $\boxed{822}$
    $297\overline{)101,871}$ $403\overline{)331,266}$

8. Challenge: One mile is 5280 feet. There are 640 acres in a square mile. How many square yards are in one acre?

    $\boxed{4,840}$

    Notes on 7: The connection is to section 9.1. In the first step of the problem on the left, for example, you know you want to multiply 297×3 because 1018 is more than 900 and much less than 1200. You can compute 297×3 in your head as $(300 - 3) \times 3 = 900 - 9 = 891$. On 8: This problem requires multiple conversions. One method is to convert one square mile to $5280 \times 5280 = 27,878,400$ square feet. Then divide by 9 to say that a square mile is 3,097,600 square yards. Then divide this by 640. (Or first convert 5280 feet to 1760 yards and get 3,097,600 by squaring 1760 which can be done with the big slide method of section 14.3, or use some other order.) The problem can also be done efficiently by thinking about prime factorizations. One mile is 1760 yards which is $16 \times 110 = 2^5\ 5\ 11$. (Miles must have been defined so a half mile, a quarter mile, an eighth of a mile, etc. were whole numbers.) So the problem is asking you to compute $(2^5\ 5\ 11)^2 \div (2^6 \times 10) = (2^{10}\ 5^2\ 11^2) \div (2^7\ 5^1) = 2^3\ 5^1\ 11^2) = 2^2 \times (2 \times 5) \times 11^2 = 4 \times 10 \times 121 = 4840$.

Name _____**Answer Key**_____

19.4 A Division Shortcut: Cancel Common Factors

1. Word Problem: Ms. Hoover's class is planning to throw eggs at Ms. Affonso's class. If Ms. Hoover brings in 14 dozen eggs and there are 24 kids in her class, how many eggs can each kid throw?

2. Cancel common factors to make each of the following division problems easier and then find the answer using long division.

   $7,200 \div 60$ $12,000 \div 24$ $165,600 \div 575$

   $\boxed{120}$ $\boxed{500}$ $\boxed{288}$

3. Xiao Yu's dog runs around their back yard 88 times. If each lap takes 15 seconds, how many minutes does the run take?

   $\boxed{22}$

4. Making Connections: What is the area in square feet of the shape shown below?

   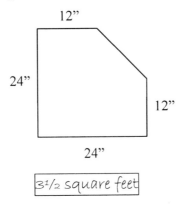

   $\boxed{3\frac{1}{2}\ \text{square feet}}$

5. Challenge: The University of Maryland has 37,037 students. In 2011 it was given a budget of $1,626,357,000. Can you think of a way to use the opposite of the method you learned in this chapter to make it easier to use long division to figure out how much the university spent per student? (Hint: What's a good way to compute $1,435 \div 5$?)

   $\boxed{\$43,911.68}$

Notes on 1: The 24 kids share 14×12 eggs so the answer is $14 \times 12 / (2 \times 12) = 7$. On 2: In the first cancel the 0's and compute $720 \div 6$. In the second cancel a factor of 12 and compute $1000 \div 2$. In the third cancel a factor of 25 and compute $6624 \div 23$. On 4: The connection is to section 10.6. Computing the area of the full square and then subtracting the area of the upper right triangle the area in square inches is $(24^2 - \frac{1}{2}\ 12^2)$. Dividing each term separately by 12^2 gives $4 - \frac{1}{2} = 3\frac{1}{2}$. An easier method is to draw in dashed horizontal and vertical lines and realize that they divide the shape into 3 one-foot-by-one-foot tiles plus a triangle that's half of a tile. On 5: Notice that $37,037 \times 3 = 111,111$. Hence, the per pupil spending can be calculated as $1,626,357,000 \times 3 \div 111,111$. It's still a bit of a pain to do without a calculator, but dividing by 111,111 is nice because each of the multiplications is easy.

Name _____ **Answer Key** _____

19.5 Converting From Base 10 to Base 8

1. Word Problem: If Maggie Simpson was born on March 18, how many days old was she be on May 17 of the same year? Give your answer in base 8.

$$74_{(8)}$$

2. Convert each number below from base 10 to base 8 by figuring out how many 64's, 8's, and 1's it takes to make the number.

$81_{(10)} = \underline{121}_{(8)}$ $\qquad$ $365_{(10)} = \underline{555}_{(8)}$

$\underline{1} \times 64 = \underline{64} \rightarrow 81 - \underline{64} = \underline{17}$ $\qquad$ $\underline{5} \times 64 = \underline{320} \rightarrow 365 - \underline{320} = \underline{45}$

$\underline{2} \times 8 = \underline{16} \rightarrow 17 - \underline{16} = \underline{1}$ $\qquad$ $\underline{5} \times 8 = \underline{40} \rightarrow 45 - \underline{40} = \underline{5}$

$\underline{1} \times 1 = \underline{1} \rightarrow 1 - \underline{1} = \underline{0}$ $\qquad$ $\underline{5} \times 1 = \underline{5} \rightarrow \underline{5} - \underline{5} = \underline{0}$

3. If the base 10 number 827 is written in base 8 what is the last digit?

$$3$$

4. Convert from base 10 to base 8 by dividing by 8 three times and writing the remainders from right to left.

$92_{(10)} = \underline{134}_{(8)}$ $\qquad$ $365_{(10)} = \underline{555}_{(8)}$ $\qquad$ $255_{(10)} = \underline{377}_{(8)}$

```
   11r4      1r3    0r1
8)92      8)11    8)1
  88        8       0
   4        3       1
```

5. Making Connections: 22 students go to a park for a field trip. Eight-fingered aliens land nearby to bring two students back to their planet. In how many different ways can they choose the two students? Write your answer in base 8.

$$347_{(8)}$$

Notes on 1: May 17 is 60 days after March 18. On 3: 827 ÷ 8 = 103 r3 so the last digit in base 8 is a 3. On 5: The connection is to section 13.4. There are 22 students they can pick first, then 21 they can pick second, but this is double-counting groups so the number of groups is 22 × 21 ÷ 2 = 11 × 21 = 231. This can be converted to base 8 by either method.

© 2013 Glenn Ellison

20.1: Simplifying Fractions

1. Word Problem: A chocolate bar is divided into 15 squares. Anna eats 10 of them. What fraction of the chocolate bar is this? Give your answer in simplest form.

$$\boxed{2/3}$$

2. In each of the fractions below the numerator and denominator have exactly one prime factor in common. Write the common factor below each fraction.

$$\frac{6}{33} \qquad \frac{5}{45} \qquad \frac{4}{34} \qquad \frac{15}{57} \qquad \frac{38}{95} \qquad \frac{187}{2890}$$

$$\underline{\ 3\ } \qquad \underline{\ 5\ } \qquad \underline{\ 2\ } \qquad \underline{\ 3\ } \qquad \underline{\ 19\ } \qquad \underline{\ 17\ }$$

3. Simplify the fractions below by canceling common factors.

$$\frac{6}{33} = \frac{2}{11} \qquad\qquad \frac{55}{135} = \frac{11}{27} \qquad\qquad \frac{45}{81} = \frac{5}{9} \qquad\qquad \frac{22}{132} = \frac{1}{6}$$

4. The fractions below are harder to simplify. See if you can simplify them by factoring the numerator or denominator and then checking to see if the factors are common factors.

$$\frac{63}{364} = \frac{9}{52} \qquad\qquad \frac{247}{3250} = \frac{19}{250} \qquad\qquad \frac{135}{999} = \frac{5}{37} \qquad\qquad \frac{704}{3080} = \frac{8}{35}$$

5. Making Connections: Add each pair of fractions and then cancel common factors to write the answers in simplest form.

$$\frac{1}{4} + \frac{5}{12} = \frac{2}{3} \qquad\qquad \frac{1}{6} + \frac{2}{15} = \frac{3}{10} \qquad\qquad \frac{5}{8} + \frac{7}{18} = \frac{73}{72}$$

6. Challenge: Simplify $\dfrac{14{,}161}{999{,}999}$.

$$\boxed{2{,}023/142{,}857}$$

Notes on 4: In 63/364 a good first step is $63 = 9 \times 7$. The denominator is not a multiple of 3 (or 9) because the sum of the digits is not a multiple of 3. It is a multiple of 7. (Recall that a 365-day year is 52 weeks plus one day.) In 247/3250 the denominator is relatively easy to factor because it's obviously a multiple of 25×10. In the third, start with $135 = 27 \times 5$ and $999 = 9 \times 111 = 9 \times 3 \times 37$. In 704/3080 a good starting point is $3080 = 10 \times 308 = 10 \times 4 \times 77$. 704 is also obviously a multiple of 4 and $704 = 4 \times 176$ makes it obvious that 11 is also a common factor. On 5: The connection is to section 17.4. In the first problem use 12 as a denominator. Use 30 or 90 in the second and 72 or 144 in the third. On 6: The denominator is a good place to start especially if you remember from section 16.5 that $1001 = 7 \times 11 \times 13$. Hence, 999,999 factors as $999 \times 1001 = (9 \times 3 \times 37) \times (7 \times 11 \times 13)$. The numerator is not a multiple of 3. It is divisible by 7: $14{,}161 = 2{,}023 \times 7$. It can't be simplified further because 2,023 is not a multiple of 11, 13 or 37.

Name _____ **Answer Key** _____

20.2: Fractions Bigger Than One: Common Fractions and Mixed Numbers

1. Word Problem: Anna is making chocolate chip cookies. The recipe calls for three-quarters of a cup of brown sugar. If Anna wants to make a triple recipe, how many cups of brown sugar will she need? (Give your answer as a mixed number.)

    $\boxed{2^{1}/_{4}}$

2. Convert the mixed numbers below to common fractions.

    $$1\frac{3}{4} = \frac{7}{4} \qquad 2\frac{1}{2} = \frac{5}{2} \qquad 5\frac{1}{6} = \frac{31}{6} \qquad 13\frac{2}{3} = \frac{41}{3}$$

    $$13\frac{3}{4} = \frac{55}{4} \qquad 12\frac{1}{12} = \frac{145}{12} \qquad 5\frac{13}{17} = \frac{98}{17} \qquad 39\frac{1}{41} = \frac{1600}{41}$$

3. Use approximate division to convert the common fractions below to mixed numbers.

    $$\frac{33}{6} = 5\frac{1}{2} \qquad \frac{117}{55} = 2\frac{7}{55} \qquad \frac{45}{14} = 3\frac{3}{14} \qquad \frac{355}{113} = 3\frac{16}{113}$$

4. Convert the common fractions below to mixed numbers.

    $$\frac{317}{6} = 52\frac{5}{6} \qquad \frac{217}{9} = 24\frac{1}{9} \qquad \frac{635}{25} = 25\frac{2}{5} \qquad \frac{2925}{175} = 16\frac{5}{7}$$

5. Which is bigger: two-thirds of $8\frac{1}{3}$ or one-half of $\frac{76}{7}$?

    $\boxed{\text{Two-thirds of } 8\frac{1}{3}}$

6. Making Connections: Find $\frac{1}{15} + \frac{2}{15} + \frac{3}{15} + \cdots + \frac{14}{15}$ as a mixed number.

    $\boxed{7}$

7. Challenge: Arda uses the digits 0, 1, 2, ..., 9 once each to make a common fraction that is NOT in simplest terms. If the numerator and the denominator are five-digit numbers, what is the largest number he could have made? Give your answer as a mixed number.

    $\boxed{98756/10234}$

Notes on 4: The last problem is easier if you first divide on top and bottom by 25. On 5: The easiest way to work with the first is to convert 8⅓ to 25/3, then multiply by 2/3, then convert back to 5 5/9. For the second first convert 76/7 to 10 6/7 then take half of that. On 6: The connection is to section 3.3. Adding pairs of numbers from the outside in gives 1+2+3+…+14 = 15×7. On 7: If you ignore that the fraction needs to be not in simplest terms the answer would be 98765/10234. But 98765 and 10234 have no common factors so we need to make the numerator smaller and/or the denominator bigger. 98756/10234 makes both numbers even by making the numerator a little smaller. This fraction is bigger than 98765/10235 so no fraction that uses a denominator other than 10234 can give a bigger answer. 98756 is the second biggest number using the digits 9,8,7,6, and 5.

Name _____ **Answer Key** _____

20.3: Canceling in Multiplication Problems

1. Word Problem: Two-thirds of the students on the Park Forest math team are girls. Three-fifths of the girls on the math team have black hair. What fraction of the students on the math team are black-haired girls?

   2/5

2. What is two-thirds of three-fifths of five-sevenths?

   2/7

3. Cancel common factors to make each of the following multiplication problems easier and then multiply.

   $$\frac{1}{2}\times\frac{2}{3}=\frac{1}{3} \qquad \frac{2}{5}\times\frac{5}{8}=\frac{1}{4} \qquad \frac{1}{7}\times\frac{14}{45}=\frac{2}{45} \qquad \frac{10}{13}\times\frac{9}{25}=\frac{18}{65}$$

   $$\frac{1}{4}\times\frac{5}{8}\times\frac{16}{17}=\frac{5}{34} \qquad \frac{4}{9}\times\frac{6}{25}\times\frac{35}{88}=\frac{7}{165} \qquad \frac{12}{115}\times\frac{9}{56}\times\frac{46}{243}=\frac{1}{315}$$

4. Making Connections: What is the area in square inches of the triangle shown below?

   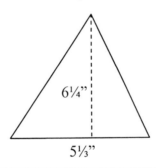

   6¼"

   5⅓"

   50/3 or 16⅔ square inches

5. Challenge: How many fifths of $\dfrac{15}{102}$ is three-sevenths of $\dfrac{14}{17}$?

   12

Notes on 4: The connection is to section 10.4. The area of a triangle is ½ × base × height. Convert the fractions to common fractions. On 5: Doing both multiplications (the "of's") makes the question. 'How many $\dfrac{3}{102}$'s is $\dfrac{6}{17}$?' Students who know how to divide fractions can compute $\dfrac{6}{17}\div\dfrac{3}{102}$. Otherwise, covert $\dfrac{6}{17}$ to a common denominator of 102 which makes it intuitive that that $\dfrac{36}{102}$ is 12 times as big as $\dfrac{3}{102}$.

HMES

© 2013 Glenn Ellison

Name _____ **Answer Key** _____

20.4: Comparing Fractions via Funny Subtraction

1. Word Problem: Twenty one of the 47 students at the Ward School scored advanced on the 4th grade math MCAS. Twenty nine of the 62 students at the Bridge School got advanced scores. At which school was the fraction of students with advanced scores larger?

    > The Bridge School

2. Use the repeated subtraction method to figure out which fraction in each pair is bigger.

    $$\frac{17}{53} > \frac{16}{51}$$ $$\frac{2}{7} < \frac{1}{3}$$ $$\frac{48}{137} < \frac{15}{41}$$ $$\frac{113}{355} > \frac{55}{173}$$

    $$\frac{11}{23} < \frac{16}{33}$$ $$\frac{98}{245} > \frac{39}{98}$$ $$\frac{355}{113} > \frac{333}{106}$$

3. Arnav has taken three tests and gotten 85, 93, and 95 points on them (all out of 100). Manisha has only taken two of the tests so far. She scored 88 and 94 on them. Who has a higher average?

    > Their averages are the same (91).

4. Making Connections: Which is larger: the probability of getting a sum of 8 when rolling two six-sided dice or the probability of having a fair coin come up Tails three times in a row?

    > The probability of getting a sum of 8

5. Challenge: Which is bigger: $\frac{111}{231}$ or $\frac{11}{23} + \left(\frac{12}{23} \times \frac{1}{230}\right)$?

    > The second fraction is bigger.

Notes on 4: The connection is to sections 18.2 and 18.3/4. The first probability is 5/36 and the second is 1/8. Compare with funny subtraction or cross-multiplication. On 5: Start by rewriting the first fraction as (111×23)/(23×231) and the second as (11×230+12)/(23×230). (Don't actually do the multiplications.) Funny subtracting the right fraction from the left makes the left one equal to 11/23. (The calculation is easier you remember that 11×230 is 110×23 and use the distributive property.) This is obviously smaller than the fraction on the right if you look at the original way that that fraction was given in the question.

20.5: Rational and Irrational Numbers

1. Word Problem: Pranav's teacher taught him that $\sqrt{p}$ is an irrational number if p is a prime number. He also taught them that the sum of an irrational number and a rational number is irrational. He then gave the class a multiple choice test and asked them to circle the rational number. What was the right answer?

   (a) $2 + \sqrt{2}$ (b) $\sqrt{5}$ (c) $\sqrt{9} + \dfrac{3}{5}$ (d) $\sqrt{8+9}$

   $\boxed{(c)\sqrt{9} + 3/5}$

2. Say whether the numbers below are rational or irrational.

   $\dfrac{2}{15}$ _____rational_____ $\sqrt{2}$ _____irrational_____

   $1 + \sqrt{2}$ _____irrational_____ $\dfrac{13}{17} + \dfrac{11}{45}$ _____rational_____

3. If a and b were integers with $a^2 = 3b^2$, which of a and b would you know must be a multiple of 3? How do you know this?

   $\boxed{\text{a would be a multiple of 3 because } 3b^2 \text{ is a multiple of 3.}}$

   (Once you know this, b must also be a multiple of 3 because a^2 is a multiple of 9.)

4. Give an example of a number x for which x^3 is irrational. Explain how you know that x^3 is irrational.

   $\boxed{\text{Many answers are possible. One example is } \sqrt{2} \text{ because } (\sqrt{2})^3 = 2\sqrt{2}.}$

5. Making Connections: Is the area of the triangle below a rational or irrational number of square inches?

   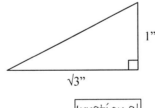

   1"

   √3"

   $\boxed{\text{Irrational}}$

6. Making Connections: Use the distributive property to help figure out whether $(1 + \sqrt{2})^2$ is rational or irrational.

   $\boxed{\text{Irrational}}$

   Notes on 5: The connection is to section 10.4. The area of a triangle is ½ × base × height.

   On 6: Connection to section 9.2 or 14.4. $(1 + \sqrt{2})^2 = 1^2 + 2 \times 1 \times \sqrt{2} + (\sqrt{2})^2 = 3 + 2\sqrt{2}$.

21.1: The Decimal System

1. Word Problem: In the first four swim meets of the year Stephanie's freestyle times were 28.21, 29.1, 28.102, and 28.11 seconds. What were her fastest and slowest times?

    Fastest: 28.102. Slowest: 29.1

2. Write each of the numbers below as a decimal.

    $134\dfrac{2{,}374}{10{,}000} =$ 134.2374 $45 + \dfrac{3}{10} + \dfrac{4}{100} + \dfrac{7}{1000} =$ 45.347

    $7\dfrac{849}{100{,}000} =$ 7.00849 $102 + \dfrac{3}{100} + \dfrac{5}{10{,}000} + \dfrac{4}{10{,}000{,}000} =$ 102.0305004

3. Write each of the decimals below as a mixed number.

    $101.632 = 101\dfrac{632}{1{,}000}$ $2{,}731.002 = 2{,}731\dfrac{2}{1{,}000}$ $31.00403217 = 31\dfrac{403{,}217}{100{,}000{,}000}$

4. Making Connections: Use the difference-of-squares formula to help find $17\dfrac{13}{100} \times 16\dfrac{87}{100}$.

    Give the answer as a decimal.

    288.9831

5. Challenge: Try to write out the complete decimal representation of the number $0 + \dfrac{1}{10} + \dfrac{2}{100} + \dfrac{3}{1{,}000} + ... + \dfrac{49}{10^{49}} + \dfrac{50}{10^{50}}$. How many times does the digit 9 appear? How many times does the digit 8 appear?

    9 appears 5 times. 8 appears 0 times.

    (The decimal is 0.12345679012345679012345679012345679012345679012340.)

Notes on 4: Connect to section 9.4. $(17 + 13/100) \times (17 - 13/100) = 17^2 - 169/10{,}000$. On 5: The best way to do this problem at this point is probably to treat it as an addition with carrying problem add 0.1234567890…1234567890 to 0.00…011…122…233…344…450. Just be careful with the carrying and remember that the second decimal has 8 0's, 10 1's, 10 2's, 10 3's, 10 4's, 1 5, and 1 0. After adding for a little while students will probably realize that (except for the last digit) there's a pattern: the digits 123456790 keep repeating (without any 8's). The reason why the pattern in present relates to the challenge problem on worksheet 21.3. Using the distributive property or long multiplication one can see that the first 49 terms match what you get when you multiply 1.111111… × 1.111111… (and then divide by 10).

HMES   © 2013 Glenn Ellison

Name _____ **Answer Key** _____

21.2: Converting Fractions to Decimals: Divide

1. Word Problem: Ms. Gao's class was making oobleck. Each student was supposed to put ¼ cup of corn starch in a bowl. The first 21 students did this, but then Ben poured the rest of the box in his bowl without measuring. If the box originally had 6.3 cups, how much did Ben pour in his bowl?

   $\boxed{1.05 \text{ cups}}$

2. Say what number you would multiply the top and bottom of each fraction by to find an equivalent fraction that had 10, 100, or 1,000 as the denominator, and then convert them to decimals using the equivalent fraction method.

   $\dfrac{3}{25}$ ×4 $\dfrac{11}{25}$ ×4 $\dfrac{101}{4}$ ×25 $\dfrac{617}{50}$ ×2 $\dfrac{3}{40}$ ×25

   $\dfrac{3}{25} = 0.12$ $\dfrac{11}{25} = \boxed{0.44}$ $\dfrac{101}{4} = \boxed{25.25}$ $\dfrac{617}{50} = \boxed{12.34}$ $\dfrac{3}{40} = \boxed{0.075}$

3. Convert each fraction to a decimal using the long-division method.

   $\dfrac{1}{4} = \boxed{0.25}$ $\dfrac{13}{4} = \boxed{3.25}$ $\dfrac{17}{25} = \boxed{0.68}$ $\dfrac{11}{8} = \boxed{1.375}$ $\dfrac{13}{400} = \boxed{0.0325}$

   $4\overline{)1.00}$ $4\overline{)13.00}$ $25\overline{)17.00}$ $8\overline{)11.000}$ $400\overline{)13.000}$

4. Challenge: Computer memory is often measured in megabytes. People think of a megabyte as a million bytes, but it is actually $2^{20} = 1,048,576$. Ms. Mao asked her class to write the fraction of a megabyte needed to store their name in memory. She meant for Kevin to write $5/1,000,000 = 0.000005$, but he likes to give exact answers so he started using long division to find $\dfrac{5}{1,048,576}$. Kate thought about this using equivalent fractions and realized it was going to take Kevin a long time. How many digits will be to the right of the decimal point when Kevin is done?

   $\boxed{20}$

   _____

   Notes on 5: Multiplying the numerator and denominator by 5^{20} shows that the fraction is equivalent to $\dfrac{5^{21}}{10^{20}}$. The decimal for $\dfrac{a}{10^b}$ has b digits after the decimal point if a isn't a multiple of 10.

Name _____**Answer Key**_____

21.3: Repeating and Terminating Decimals

1. Word Problem: Angela was born in China. She moved to the U.S. on her 7th birthday. On her 12th birthday, what portion of her life had been spent in the U.S.? Give your answer as a decimal.

    $$\boxed{0.41666...}$$

    (Answers to questions 1, 2, 5, 6 can be written using bar notation or with "...".)

2. Use long division to write each of the fractions below as a decimal.

    $\frac{2}{3} = \boxed{0.666...}$ $\frac{5}{6} = \boxed{0.8333...}$ $\frac{3}{11} = \boxed{0.2727...}$ $\frac{5}{18} = \boxed{0.2777...}$

    $3\overline{)2.0000}$ $6\overline{)5.0000}$ $11\overline{)3.0000}$ $18\overline{)5.0000}$

3. Write each of the decimals below using the bar notation for repeating decimals.

    $0.41666... = 0.41\overline{6}$ $3.2727... = 3.\overline{27}$ $4.5060606... = 4.5\overline{06}$

4. Say whether the decimal expansion of each fraction below will be terminating.

    $\frac{1}{13}$ ___No___ $\frac{13}{4}$ ___Yes___ $\frac{7}{11}$ ___No___ $\frac{3}{6}$ ___No___

    $\frac{17}{40}$ ___Yes___ $\frac{3}{24}$ ___No___ $\frac{5}{64}$ ___Yes___ $\frac{21}{262}$ ___No___

5. Making Connections: What is the probability of rolling two numbers that add up to 4 when you roll two six-sided dice? Give your answer as a decimal.

    $$\boxed{0.08333...}$$

6. Challenge: Write $\frac{10}{81}$ as a decimal.

    $$\boxed{0.123456790123456790...}$$

Notes on 5: The connection is to section 18.2. The probability is 3/36 = 1/12. On 6: A good way to do this is just to use long division on $81\overline{)10.0000...}$. This decimal looks like the answer to the challenge problem on worksheet 21.1 because 10/81= (10/9) × (10/9) = 1.111... × 1.111... = $1 + \frac{2}{10} + \frac{3}{100} + \frac{4}{1000} + \cdots$. (Use long multiplication to see this.)

21.4: Playing with Sevenths

1. Word Problem: Krishna tried computing one-seventh using long division, but he didn't know about repeating decimals, so he just kept dividing until class ended. If he wrote 25 digits after the decimal point, what was the last digit he wrote?

2. Use long division to write three-sevenths as a decimal.

   $0.428571\ 428571...$

3. Using the result of the previous question write $\dfrac{300}{7} = 42\dfrac{6}{7}$ as a decimal.

   $42.857142\ 857142...$

4. The decimal for $\dfrac{1}{27}$ is $0.037037037...$. Find the decimals for $\dfrac{2}{27}$ and $\dfrac{3}{27}$?

   $2/27 = 0.074074...$ $3/27 = 0.111111...$

5. Making Connections: The decimal for $\dfrac{1}{2439}$ is $0.0004100041...$ Use the distributive property to find the decimal for $\dfrac{1001}{2439}$.

   $0.41041\ 41041...$

6. Challenge: Find the decimal for $\dfrac{1}{7} + \dfrac{1}{11} + \dfrac{1}{27}$.

   $0.270803\ 270803\ 270803...$

Notes on 5: The connection is to section 9.1. $1001 \times \dfrac{1}{2439} = 1000 \times \dfrac{1}{2439} + 1 \times \dfrac{1}{2439}$. Then use addition on the decimals $0.4100041000... + 0.0004100041....$ On 6: Write each of the three fractions as a repeating decimal and use addition with carrying to compute $0.142857... + 0.090909... + 0.037037...$ The one tricky aspect of addition with carrying with nonterminating decimal is that you still need to go from right to left, but you can just start in any place where you know there's no carrying and everything will be OK to the left of that. Here, the form of the three decimals makes it clear that it will repeat every six digits.

21.5: A Shortcut: Memorize the Answers

1. Word Problem: Amy ate three-eighths of a chocolate bar. Chloe ate 0.3636… of an identical chocolate bar. Who ate more?

   Amy

2. Give the decimal equivalent for each of the fractions below.

$\frac{1}{8} = 0.125$ $\frac{1}{4} = 0.25$ $\frac{3}{8} = 0.375$ $\frac{1}{2} = 0.5$ $\frac{5}{8} = 0.625$ $\frac{3}{4} = 0.75$ $\frac{7}{8} = 0.875$

$\frac{1}{9} = 0.\overline{1}$ $\frac{4}{9} = 0.\overline{4}$ $\frac{7}{9} = 0.\overline{7}$ $\frac{1}{6} = 0.1\overline{6}$ $\frac{5}{6} = 0.8\overline{3}$ $\frac{1}{6} = 0.1\overline{6}$ $\frac{5}{6} = 0.8\overline{3}$

$\frac{1}{11} = 0.\overline{09}$ $\frac{2}{11} = 0.\overline{18}$ $\frac{5}{11} = 0.\overline{45}$ $\frac{3}{11} = 0.\overline{27}$ $\frac{1}{12} = 0.08\overline{3}$ $\frac{7}{12} = 0.58\overline{3}$ $\frac{5}{12} = 0.41\overline{6}$

3. Give a simplified fraction equivalent to each decimal.

$0.25 = \frac{1}{4}$ $0.375 = \frac{3}{8}$ $0.4 = \frac{2}{5}$ $0.6 = \frac{3}{5}$ $0.75 = \frac{3}{4}$ $0.8 = \frac{4}{5}$

$0.\overline{09} = \frac{1}{11}$ $0.625 = \frac{5}{8}$ $0.\overline{44} = \frac{4}{9}$ $0.125 = \frac{1}{8}$ $0.1\overline{6} = \frac{1}{6}$ $0.4 = \frac{2}{5}$

$0.08\overline{3} = \frac{1}{12}$ $0.41\overline{6} = \frac{5}{12}$ $0.\overline{36} = \frac{4}{11}$ $0.875 = \frac{7}{8}$ $0.\overline{54} = \frac{6}{11}$ $0.7 = \frac{7}{10}$

4. Making Connections: Find $0.\overline{27} + 0.1\overline{6} + (0.\overline{6} \times 0.\overline{09})$.

   1/2

5. Challenge: The decimal expansions for $\frac{5}{6}$ and $\frac{1}{12}$ are $0.8\overline{3}$ and $0.08\overline{3}$. Can you find another pair of fractions in lowest terms that are less than one and have denominators less than twenty that also have decimal expansions that differ in this way (i.e. with one decimal consisting of a zero and then the same digits that the other one has after the decimal point)? How many such pairs are there?

   There are eight more pairs: (2/3, 1/15), (5/7, 1/14), (5/8, 1/16), (5/9, 1/18), (10/11, 1/11), (10/13, 1/13), (10/17, 1/17), (10/19, 1/19)

Notes on 4: The connection is to section 17.4. Convert the decimals to fractions (3/11, 1/6, and 2/3 × 1/11) and then add the fractions. 66 is a common denominator. On 5: The second fraction in each pair must be one-tenth of the first. 1/10 × a/b won't have a denominator less than 20 unless a is a multiple of 2 (with b less than 4), or a is a multiple of 5 (with b less than 10), or a is a multiple of 10 (with b less than 20).

21.6: What is 0.999999...?

1. Word Problem: Ms. Bai asked her class to find the digit that would be in the one-thousandths place when the fraction $\frac{7}{50}$ was written as a decimal. She expected everyone to say 'zero', but Caira knew that there was another answer that is also possible. What is the other answer?

    9

2. Write standard terminating decimals equal to each of the decimals below.

   0.999... = 1   0.1999... = 0.2 $3.13\overline{9}$ = 3.14

3. Evaluate $0.\overline{21} \times 99$ by computing $0.\overline{21} \times (100 - 1)$. Use this result to find a simplified fraction that is equal to $0.\overline{21}$.

   7/33

4. What simplified fraction that is equal to $0.2\overline{5}$? (Hint: Multiply by 100 and by 10.)

   23/90

5. Making Connections: Find 0.1212... + 0.5151... + 0.2121... + 0.1515...

   1

6. Challenge: In a society that used a base 8 number system, numbers that are less than one would probably be written using an "octimal" system in which the place values are one-eighth, one sixty-fourth, etc. What whole number is equivalent to $0.\overline{7}_{(8)}$? What fraction is equivalent to $0.\overline{21}_{(8)}$?

   $1_{(8)}$ and $21_{(8)}/77_{(8)}$ (or 1 and 17/63 if one answers in base 10)

Notes on 1: The fraction can be written as 0.14 or as 0.13999... On 5: The connection is to section 1.3. Just line up the numbers on the decimal point and use addition with carrying. (You can also reorder and add the first and third, then the second and fourth decimals if you want to do two at a time.) On 6: The number $0.777..._{(8)}$ is just like 0.999... in base 10. To convert the second problem note that in base 8 multiplying by $100_{(8)}$ which we call 64 moves the decimal point over by two places. So multiplying by 64 and then subtracting the original number gives that 63 times the number is equal to $21_{(8)}$ which we call 17.